The Constant Fire

The Constant Fire

Beyond the Science vs. Religion Debate

Adam Frank

UNIVERSITY OF CALIFORNIA PRESS

Berkeley   Los Angeles   London

Every effort has been made to identify and locate the
rightful copyright holders of all material not specifically
commissioned for use in this publication and to secure
permission, where applicable, for reuse of all such material.
Credit, if and as available, has been provided for all borrowed material either on-page, on the copyright page, or in
an acknowledgment section of the book. Errors, omissions,
or failure to obtain authorization with respect to material
copyrighted by other sources has been either unavoidable
or unintentional. The author and publisher welcome any
information that would allow them to correct future
reprints.

University of California Press, one of the most distinguished university presses in the United States, enriches
lives around the world by advancing scholarship in the
humanities, social sciences, and natural sciences. Its
activities are supported by the UC Press Foundation
and by philanthropic contributions from individuals and
institutions. For more information, visit www.ucpress.edu.

University of California Press
Berkeley and Los Angeles, California

University of California Press, Ltd.
London, England

Library of Congress Cataloging-in-Publication Data

Frank, Adam, 1962–.
    The constant fire : beyond the science vs. religion debate
/ Adam Frank.
        p.   cm.
    Includes bibliographical references and index.
    ISBN 978-0-520-26586-8 (pbk : alk. paper)
    1. Religion and science — History.   I. Title.
    BL245.F73   2009
    201'.65 — dc22   2008025402

Manufactured in the United States of America

17   16   15   14   13   12   11   10
10   9   8   7   6   5   4   3   2   1

This book is printed on Natures Book, which contains 50%
post-consumer waste and meets the minimum requirements
of ANSI/NISO z39.48–1992 (R 1997) (*Permanence of Paper*).

To my children,
Sadie Ava Frank and Harrison David Frank,
for their boundless courage
and unfailing sense of humor

CONTENTS

ACKNOWLEDGMENTS

This work is the fruit of many conversations with many wonderful friends and colleagues. I would not have gained the understanding reflected in the words that follow without their intelligence, interest, and long-suffering patience. First and foremost, I am indebted to K C Cole. She was my first editor and has been a mentor and friend for more than ten years. I must also thank Paul Green and Robert Pincus for two decades of splitting hairs over everything from the nature of Quality to whether the Who were more significant than the Beatles (they were, OK). I am indebted to their long years of intellectual friendship. Conversations with Margaret King and Adam Turner led me, over the years, in a variety of directions for which I am most grateful. I must also acknowledge many conversations with the author Rafe Martin, who led me to the title of this book, which comes from a poem by Wallace Stevens. He has helped enormously with the mythologies that are included at the end of each chapter. Rafe's inspiration and deep knowledge of the role of narrative in human experience were a boon to me, and I am truly grateful for his help and guidance. Along the same lines, my coffee-time talks with Steve Carpenter, one of the finest painters I have had the pleasure to know, were a constant source of inspiration. I came away from every meeting feeling like I had just bungee-jumped off a cliff of great ideas.

Eric Blackman, my longtime collaborator in all things astrophysical, served as my principal skeptic in the development of these ideas. I am grateful for his open-mindedness on these issues. I would also like to thank my other colleagues, Den Watson, Alice Quillen, Bill Forrest, and Judy Pipher, for their friendship and forbearance during the writing. My thesis adviser, Bruce Balick, has always been a tremendous inspiration to me, and he was a great pleasure to talk with during the writing of this book. I must also thank Vincent Icke for his instruction as a scientist. I would also like to thank Tom Jones at the University of Minnesota for the fine example of what science with compassion and interest looks like.

This book grew out of a Bridging Fellowship offered to faculty by the University of Rochester, and I am forever grateful to that wonderful home institution for supporting such efforts. A number of faculty members at the University of Rochester Department of Religion and Classics were generous with their time and ideas. I wish to thank Dean William Green for some of the first directions this book took through conversations and initial readings. Emil Homerin was generous with his time, and I am thankful for his guidance. Monica Florence gave me a great deal of help with issues in mythology. I am also grateful to Wendy Doniger and Adam Shapiro for their input on these ideas. Ursula Goodenough provided me with a number of key ideas and was enthusiastic in her encouragement. I am also indebted to Piet Hut, who remains an inspiration to me for his broad ideas and tireless efforts in these domains.

Tom and Mary Slothower proved to be good and dear friends during the writing of this book, and I am grateful for their encouragement and companionship. Rebecca Gilbert inspired me with her music, creativity, and enduring friendship. Along the same lines, Scott and Katie MacDonald contributed greatly to my state of mind in the long months of writing, and I am happy for their friendship. Donnita George was a great inspiration and friend when I needed it most. I also benefited from many wonderful talks by Jerome Laufer on science, mythology, the best bullet mics for blues harp, and kayak combat rolls.

Writing has been a moonlighting activity for me over the years, and it has been wonderful to have the chance to work with some great magazines and editors during the past ten years. I am very grateful to Corey Powel at *Discover* who has taught me a great deal and David Eichler at *Astronomy* magazine for his input on many stories I have written for him. I would like to thank *Tricycle* magazine and Ian Collins for giving me the chance to come out of the closet as a scientist with spiritual inclinations. I would also like to thank Robin Jaeckel for her constant friendship and support throughout the writing of this book. She has often helped me keep perspective in difficult times.

Finally and most important, I wish to thank my family. My children; my mom, Ingrid Frank; my stepfather, George Richardson; and my sister, Elisabeth, and brother-in-law, Hendrik Helmer, are all heroes in the narrative of this particular life. I would not be complete if I did not thank Holly Merrill. And, of course, my brother, David, and father, Jerry. Thank you.

House of the Rising Sun

Newgrange, Ireland. The entrance was barely wide enough for a grown man to pass through. Unlike the bright Irish summer day behind us, the path ahead was dark, cool, and damp. The ceiling hung low, and the walls pressed in close. We threaded single file down a narrow passageway. The stone walls around us had been set in place a full half millennium before the pyramids were finished. The passageway opened into a low circular chamber set into the stone. As my Irish astronomer colleagues and I circled the chamber to make room for our guide, the silence of five millennia settled on us like a layer of ash.

"One morning each year," our guide intoned in a rich baritone, "on the day of the winter solstice, the rising sun aligns with the passageway you just traversed. On that morning, and a few others before and after, a beam of sunlight pierces the twenty-four-meter-long hallway and falls here in the central chamber. The light within this black chamber flares for a mere fourteen minutes. Then it passes. The sun rises above the hill, the alignment is broken, and the chamber remains dark for another year. What rituals and practices were carried out here in that brief period of illumination we do not know."[1]

This day trip to Newgrange, a Neolithic monument fifty miles north of Dublin,[2] had been arranged by Tom Ray, my colleague from the Dublin Institute for Advanced Studies. Ray is often called on to verify the astronomical orientation of Ireland's many archaeological sites, and he had recently completed work at Newgrange. I was in Ireland for a week as a visiting astronomer at the institute, working with Ray and his team on theories of star formation. While we were touring Newgrange, racks of supercomputers were hard at work back at the institute, stirring the mix of digital abstractions we had launched that morning into a simulation of a newborn star. By the next day the simulation would be done.

Now, standing in the center of one of the oldest human edifices in the world, I was simultaneously lifted out of the day-to-day concerns of a working scientist and returned to the very root of that science. The builders of the massive quartz-lined monument clearly understood the principles of fundamental astronomy. Their keen intelligence and acute powers of astronomical observation were evident throughout the site. But to imagine the builders of Newgrange as only Neolithic proto-astronomers would be to miss the experience of the central chamber. The tremendous effort required to construct this site was not meant to simply create some kind of large calendar or calculator (many-ton slabs of stone had to be dragged miles without the aid of horse or wheel). Here metaphors became concrete. The dark chamber was the world's womb, a sacred space where once each year the promise of light and life was renewed. Here the close observation of the world merged with a profound reverence for it, and meaning, order, and purpose emerged. Newgrange was a place of myth where stories of time, space, and origins were given physical form. Here narratives of recurrent cycles and creation were raised to a higher purpose, evoking the world's sacred character in all who participated, all who heard and experienced the stories. In that sense, it seemed to me that it was no different from the institute I would return to the next day.

WHERE I'M COMING FROM

Rochester, New York. I am a card-carrying believer, an unreconstructed devotee, and an unapologetic disciple. I am firm in my faith, devoted, evangelical even — the whole nine yards. What I believe in, however, is not a religion but the path and practice of science. I am a devotee of its methods and its means. I am a disciple of the universe it reveals in all its power and grandeur. I see science as one of the supreme achievements of human culture, a measure of the best we are capable of and the best we can aspire to. My own experience of the world has been deepened, made richer and more fecund through the lens of scientific practice and the insights it offers. That irreducible personal experience serves as the motivation for this book, for it has been through science that I have gained a sense of the world's immense spiritual dimension, that I have come to my most potent sense of that character of life that can only be called sacred.

I fell in love with astronomy when I was five years old. Late one night in the family library, I found the keys to the universe sketched out on the covers of my dad's pulp science-fiction magazines. From astronauts bounding across the jagged frontiers of alien worlds to starships rising discovery-ward on pillars of fire, the boundless world of possibilities revealed on those covers became the one I was determined to inhabit. Later, my love for astronomy transformed into a passion for the practice of science itself. That passion was kindled when my father's simple explanation of electric currents and sound waves transformed the terror of a booming thunderstorm into the marvel of the world's invisible structures manifesting in material form.

It is with some trepidation, then, that I began a book on the subject of science and religion. These are waters entered into with no small degree of peril. The literature is vast, and the subject, of course, touches on the deepest questions humans can ask. It is also a subject guaranteed to infuriate someone in your audience in ways that can lead to threatening letters and flaming bags of dog poop thrown on your porch. In the first years of the twenty-first century Science and Religion often evoke antagonism and very strong feelings.

It is easy to make sweeping, often unsupportable claims about fundamental antagonisms or connections between science and religion. Affirming prejudices on either side requires little thought or scholarship. A thoughtful, closely reasoned investigation that might reveal a different perspective, however, requires a mastery of many subjects, from theology to neurology. In the face of that endeavor there is a strong impulse to just let the subject lie where it is. Why even try when it is so easy for the reach of your claims to exceed the grasp of your argument?

I cannot claim mastery of most or even many of the subjects that inform the intersections between science and religion. I am a practicing scientist who responds deeply to the magnificent beauty revealed in the worlds of science *and* art *and* many of the diverse forms of spiritual endeavor. That interior response has led me to read widely in philosophy, religious studies, and mythology. My readings have by no means been exhaustive or complete. They represent explorations, an attempt to feel out a safe passage that can bridge the vast gap between the externally directed universe of science and the interior life of spiritual aspiration. My hope is that in these early explorations, incomplete as they are, I have found something useful.

It was important to begin this project with an understanding of the barriers both real and imagined that work against it. A scientist's natural inclination is to require a rational, empirically based approach complete with stringent definitions. Scientists are taught to always ask one fundamental question: Where are the data? The domains of human spiritual experience are not amenable to such interrogation. Spiritual endeavor and spiritual experience constitute a gray area perhaps connected to the physical world of science, or perhaps not. This creates a barrier to the discussion of science and religion because we scientists are careful to police the borders of our respective fields, keeping them safe from faulty reasoning and suspect conclusions that cannot stand up to our cherished standards of inquiry. A lot of the discussion about science and religion appears to us to enter those border regions. It is not clear that anything worthwhile can be said about the two. Worse, what can be said is too often irrelevant to our core questions or our means to answer those questions.

Beyond these legitimate concerns, however, science as an institution has its own prejudices about this subject. There are places we are subtly taught it is best not to go. We may run the risk of being labeled "soft" or "mushy" in our thinking if we take too positive a view of the domains of religious feeling. One must be careful about keeping one's affinities for terms like *sacred* and *spirituality* in the closet. You don't very often see scientists admitting to colleagues their deep and abiding sense of the world's inner life. It would be like breaking into tears in seventh-grade gym class after getting hit upside the head with the basketball.[3]

At this moment in history, when we face such obvious and overwhelming dangers, I believe it is time to put those prejudices aside. In a century already haunted by the specter of climate change, resource depletion, and a variety of terrorisms we need any and all tools of wisdom we can get. I write this book with a sense of urgency that the world our kids will grow up in may look very different, and not better, from the one we stumbled into. Science and religion, at their best, can both be sources of wisdom. At their worst . . . well, we all know what that looks like.

SCIENCE, RELIGION, AND EXPERIENCE

This book is not about science and religion per se; it is about science and human spiritual endeavor. It is about the human aspiration to find what is true, what is real, and to then build lives in accord with that understanding. This aspiration, this "constant fire," as I call it, is as old as humanity itself. It is, I believe, the right place to locate a different, more sustaining perspective on science and spiritual endeavor. My goal in this book is to lay out an argument that the two great enterprises arise from a common impulse. At their best they can serve a common purpose and gain their efficacy from a common source.

I am not a big fan of the word *religion,* and in this book I do my best to avoid it when I can. (I spend some time working out definitions for words like *sacred* and *spiritual* later on.) For many people the word *religion* conjures up images of institutions of power and real estate: churches, raiments, crusades, and jihads. On a personal level I have sympathy with those who can recall childhood experiences of formal religion as a long encounter with unreasoned scriptures, dogma, and empty ritual that in the end became nothing more than a demand to believe or else. Formal religions can, without doubt, be a means for people to gain access to their own interior depths. But it often seems that, at least for some fraction of the time, people have to perform end runs within their institutions to find their personal paths.

In my search for a different perspective on science and religion, I have taken as my starting point the views of William James, the American psychologist and philosopher whose work *The Varieties of Religious Experience* remains vivid and relevant one hundred years after its publication. James focused not on canon, scripture, heresy, or dogma but on the fundamental experience of spirituality erupting, in one form or another, into people's lives. This he took as the core of religious life, as opposed to the institutions of religion. Many people "practice" religion for a lifetime, sitting in church or temple once a week, without ever manifesting an elemental effort at understanding. I would *not* call this spiritual endeavor. In James's words such a person has "his religion . . . made for him by others, communicated to him by tradition, determined to him by fixed forms by imitation, and retained by habit."[4]

Using more than two hundred personal accounts gathered from a variety of sources, James detailed the ways people become aware of a spiritual domain in their lives. For some it is a fleeting moment that dims in time. For others it is a profound sense of the world's sacred character that shapes a lifelong aspiration to understand and know the source of that experience. James's perspective is crucial because it separates religious institutions from the more elemental, more personal domains of what I call spiritual endeavor. It is the latter that I target as the appropriate and critical point of contact with science and that is, I hope, a facet of the phenomenon of religion that scientists can be comfortable with and respect. What matters to me is that untold numbers of human beings have had, and continue to have, experiences of spiritual truth. The world is viewed as sacred in their lives, and they are transformed by that experience.

Spiritual endeavor often begins with a direct encounter with suffering or rapture. That experience drives a person out of the confines of self, as is explicit in the story of Siddhartha Gautama, the Indian prince who became the Buddha, or Saint Teresa of Avila, the Catholic contemplative. From there a commitment is established to explore directly matters of birth and death, the True and the Real. Experience is the seed of aspiration, the deeply rooted commitment to know. That aspiration propels one into the difficult and transformative realms of spiritual endeavor. For me these realms of experience, often forming the living heart of any particular religious tradition, are the places where connections with science can be found that are of interest, that are fruitful, and that are very much needed. I begin from the premise that there is content in those experiences. There is something, some truth discovered, that is more than simple neurochemistry gone amok.

I begin with this premise because throughout my life I have had such experiences. In a comic twist of fate, it has often been the apprehension of the world through the lens of science that became my gateway for this sense of the world's sacred character. I first felt it as a boy lying on the roof of my parents industrial New Jersey home. Staring up at the ten or so stars that could be discerned in the hazy bowl of the sky (it was New Jersey after all) the sense of the world as sacred appeared to me as the presence of indefinable beauty and perfection. Later I felt it in the science I learned — for example, the giddy vertigo of recursion relations in the mathematics of electromagnetic fields. So often the encounter with the fruits of science has brought this great and abiding sense of mystery to me. I have felt it in observatories and laboratories and classrooms. I felt it that day in Newgrange, confronted with its long-vanished architects' close observation of the sun's path across the year. I feel a lot of sympathy with the accounts I read in James's *Varieties of Religious Experience*.

The power of science is for me also based on experience. It is a way of approaching and, most important, encountering the world. As E. O. Wilson wrote, "Science is neither a philosophy nor a belief system."[5] I am interested in how the kind of experience described by James and the effort that flows from it, considered by definition "religious," connects to this other great wellspring of human creativity called science.

Language can be the most formidable barrier to a new perspective on science and religion. In my search for a vocabulary that can transcend the debates of the past I have tried to look for words that can be unpacked in fresh and innovative ways. Like William James, I am interested in what is common to the experience so many of us have of the world illuminated, made clearer and more startling. *Spiritual* is a term favored by many people uncomfortable with the specific creeds of specific religions. I use this word at times, but for a scientist it has its problems, conjuring up the root term *spirit* as an eternal transcendent entity. Another word works better for the task ahead. From my reading I have come to lean on the term *sacred*, for a number of reasons described later. Many people, scientists in particular, may be put off to hear the words *sacred* and *science* used together. It seems too New Age-ish, hinting of crystal energy channeling or other silliness. I empathize with them. I intend, however, to use the word as some scholars of religion have provided. There is a constellation of ideas in the term *sacred* that cuts across specific religions and reaches to the heart of those experiences we call "religious." The world takes on a distinctive sense when we feel its presence in-and-of-itself. When standing under the mottled band of the Milky Way or watching the power of a swift, wide river we can sense the sheer weight of the world's simple unexpected presence that is beautifully humbling. At those moments the world itself can seem full, suffused, in a word, sacred. But we must be careful not to let *sacred* become *the sacred*, another name for a preexisting reality, a thing in itself, a substitute for God. In using this word *sacred*, I am pointing explicitly to the character of the experience. I am speaking of the richly felt illumination we sense in those moments. What, if anything, lies beyond it is a subject of speculation that too easily becomes metaphysics. This is not ground that can be used to build bridges between science and spiritual endeavor. I hope to show that experience of the world's sacred character alone can be strong enough to support that weight.

BEYOND WARFARE

The traditional debate between science and religion misses the point almost entirely. In the popular media we continually see religious fundamentalists pitting their scriptural stories against the narratives of science. Here the great battle begins, and here we are all the losers. Scripture can never trump science's ability to respond to the nature of the physical world. Worse, the argument never touches the living heart of the issue. Science and spiritual endeavor are both responses to the lived sense of the world's great mystery. What matters most is not the latest results of evolutionary theory or cosmology but the common aspiration they share.

Theologically based arguments about what God can or cannot do in light of the laws of nature also miss the point. These arguments seem sterile, touching neither the lived truth of spiritual experience nor the spectacular successes of science. The urgencies of institutions, Scriptures, and theologies are explicitly not what I am interested in. This book is actually atheistic or at least nontheistic. I am not interested in theism, *ideas* about God, but in the profound experience of the world as sacred, the knowledge that flows from that experience, and, most important, its relationship to the knowledge gained from science.

Alternative approaches to science and religion based on "Eastern" perspectives are equally flawed. Claims that Buddhist worldviews are confirmed at the frontiers of quantum physics may generate press and controversy, but they too miss a simple, deeper, and more fertile truth. Like the young-Earth and intelligent design debates, much of what has been claimed leaves practicing physicists like me breathing into a paper bag to stay calm. Worse, some works from this perspective, such as Deepak Chopra's infuriating writings and the infamous movie *What the Bleep Do We Know*, substitute wishful thinking for an honest assessment of any real parallelism between science and human spiritual endeavor. Coming to a deep understanding about the structure of the world will, by its very nature, require immensely hard work.[6] This is true of science and authentic spiritual endeavor. No expensive weekend seminar on "quantum healing" or the promise of instant enlightenment will do the trick.

There is danger residing on the other side of the fence as well. We scientists run the risk of being just as prejudiced and biased as those we condemn in their dismissal or disregard for science. While the term *religion* is often colored by the taint of power and politics, the lived experience of a spiritual dimension in human life is so common that to ignore it is simply foolish. Yet there continues to be an unspoken dismissal of these perspectives in the institution of science. Those who experience spirituality in their lives and are also touched by the beauty and power of science are caught between the dogmas of both sides in the traditional debate between science and religion. They are open to seeing their own religious traditions as part of a worldwide continuum of spiritual longing. Most important, they know their way is not the only way. It is this reality that the institutions of science appear to deny just as the institutions of religion so often police the borders of belief.

I am hopeful that it is possible to go beyond these narrow views. I believe that it is possible and *necessary* to see science and human spiritual endeavor from a different perspective, one that honors both great fields of experience and effort. An alternative perspective would acknowledge the truths each can reveal and avoid jealous comparison of those truths. *In writing this book I am interested in the source or aspiration of both science and spiritual longing. This is what I call the constant fire.* I argue that this aspiration is ancient, stretching back five thousand years before we stepped into history. The place to ground such an alternative perspective is, I believe, in the great traditions and narratives of mythology.

WORLDS OF MYTH, ARCHETYPE, AND SCIENCE

Myth and science have much in common for both are the realm of story. The universe of myth is the space where humanity imaginatively represents the world and its discovered truth through sacred narratives. As the great scholar of religion Mircea Eliade put it, "Through myth the world can be apprehended as a perfectly articulated, intelligible and significant Cosmos."[7] If we deepen our understanding of myth we will see that it provides a language through which science's potential as a gateway to apprehending the world's sacred character can be opened.

The common understanding of myth is as "false story." This is an egregious error that becomes apparent when we take the long view of cultural history. The particular importance of mythic narratives is their ability to transmit essential truths within and across human cultures with poetic economy. The Swiss psychologist Carl Jung developed the concept of archetypes, or fundamental forms, that are embedded in human consciousness. They appear in individual dreams and in potent symbols that cross cultural and historical barriers. Jung developed this concept with the idea that the source of these archetypes can be found in the realm of myth. Of the many ways scholars have viewed myth, this paradigm of universality, whether or not it uses archetypes, can serve as the critical pivot point for bringing science and spiritual endeavor into an active and complementary parallel.

The idea is relatively simple, but its implications are profound. There are a set of elemental forms or ideas expressed in myth that recur in almost every culture. These stories — the hero's journey, the apocalyptic flood, the cycles of cosmic creation and destruction — express universal forms of human truth. It is no surprise, then, that we find them reappearing in cultures as different as those of the Australian Aborigines and the Viking Norsemen.

Many people are familiar with these ideas through the popular work of Joseph Campbell. For Campbell the world's true spiritual meaning was expressed through myth. In his view mythologies are narratives that convey the archetypes of human experience. Organized religion was, for Campbell, only a kind of clothing we use to cover our spirituality. Myth is its naked core. Campbell remains a controversial figure in the study of mythology, and it is not clear how far his ideas can be taken. Still the essential concept of a commonality of mythic elements across human cultures is part of a spectrum of ideas about myth that can help to establish its role as a bridge between science and religion. There are other ideas from the rich study of mythology that need to be tapped as well. Taken together, the different perspectives reveal how primordial encounters with the sacred are transformed through myth into narratives reflecting humanity's role as creative observers. In this way an understanding of myth can help bridge the scientific and modern religious perspectives.

OLD STORIES, NEW DEMANDS

The debate between science and religion is more than four centuries old. What is new is that it is now colored by an especially harsh imperative. Wisdom and understanding can no longer be separated on a planet pushed by technology and science to the limits of its carrying capacity.

Our world is saturated with the fruits and poisons of science. There can be no doubt that in spite of the grandeur and blessing of these fruits, the poisons now threaten both the human habitability of the planet and our "project of civilization."

In her great work *A Distant Mirror* the historian Barbara Tuchman looked at events of an older era, the fourteenth century, and described it as "a bad time for humanity."[8] More recently, in *The Final Hour*, Sir Martin Rees, Astronomer Royal of England, gave humanity even odds for making it out of the next century. It is difficult to look at our moment in history and not wonder about the long-term viability of what we now call civilization. Something, some planetary ethic is urgently needed to guide us through this dangerous moment in our evolution. The biologist Ursula Goodenough puts it plainly: "That we need a planetary ethic is so obvious that I need list but a few words: climate, ethnic cleansing, fossil fuels, habitat preservation, human rights, hunger, infectious disease, nuclear weapons, oceans, ozone layer, pollution, population."[9] A lot more will be required of our children than was asked of my generation. In that light I believe there is an imperative to reimagine the links between science and spiritual endeavor.

We must find the will to create a language for science and religion that transcends debates over the latest research results and focuses instead on their original and common roots. Only together can these two great human enterprises represent our best route to a new form of right understanding. Only together can they help us navigate what may be hard times ahead. Given the pressures we face as a species, we must draw on both science and religion as a source of deeper understanding about our collective place in the unfolding of time, and this has to translate into wise and skillful action.

By recognizing science as a means of manifesting what we apprehend as sacred, we can no longer see it as simply a means to an end. By understanding science as practice saturated with a deeper aspiration, we can no longer make it simply a tool for the control of the natural world. By locating science in the mythic field of spiritual endeavor, it becomes an instrument for gaining both knowledge and wisdom. That would represent a fundamental change.

THE DUBLIN INSTITUTE
FOR ADVANCED STUDIES

The space between the towering racks of high-performance computers was barely wide enough for a grown man to pass through. The room was dark, cool, and dry. I threaded between the racks, making my way by the pale illumination of muted green LED displays on the face of each computer box. A slight air-conditioned breeze blew across my face. It was an artificial wind designed to extract the heat generated by all the machines working at once.

I was in the Cluster Room of the Dublin Institute for Advanced Studies. Here, in a quiet broken only by the hum of small fans, a multitude of linked computers were silently working. They were machines meditating in the Platonic realm of mathematics. Like performers in an intricate digital orchestra, the boxes were working in parallel, tied together to create a single supercomputer pushing forward the simulation we had launched yesterday before we left for Newgrange. It would take just one more day for the equations describing gravity, gas motions, radiation, and rotation to be fully digitally transformed and provide my colleagues and me with a virtual glimpse of star formation.

As I stood in the midst of the machines, in the quiet and the coolness and the muted light, I recognized that I was once again in a place where metaphors became concrete. Like the central chamber of Newgrange, the close observation of the world merged here with a profound reverence for it. There was genius made manifest in those boxes and their tangle of interconnecting wires. There was so much effort, time, and wealth brought to bear in their creation. In their midst, surrounded by a cloud of mathematical abstraction slowly being transformed into a story of star birth, I felt two worlds slip into a parallel, complementary affinity like two magnets orienting themselves to their respective lines of force. This room was also a place of myth and science, where apprehension of the illuminated world moved from feeling into form. A sense of presence hung in the air for a moment, and I lingered with it. I knew this sense would pass. I also knew it would return — to me, to my colleagues, to the students I would lecture to that fall, to all of us. It was the voice of the world speaking in its own language. It would return because it has always been with us. I waited another moment and then went back to work.

PART I

The Map

Reimagining Science, Myth, and the Sacred

CHAPTER I

The Roots of Conflict

Science and Religion before Divorce

The Religion that is afraid of Science dishonors God and commits suicide.
Ralph Waldo Emerson, *Journal* (1831)

There may be a great fire in our soul, yet no one ever comes to warm himself at it, and the passers-by see only a wisp of smoke.
Vincent van Gogh,
quoted in Cliff Edwards, *The Shoes of Van Gogh*

Great men are meteors designed to burn so that the earth may be lighted.
Napoleon Bonaparte,
quoted in Richard Alan Krieger, *Life's Ideal*

It is winter in Rome. The new year and the new century, 1600, are little more than a month old. In the Campo di Fiori, a popular square near the heart of the city, a crowd gathers, anxiously awaiting the spectacle of an execution. Today Giordano Bruno, philosopher, astronomer, and former Dominican monk, will be burned at the stake. People stand tiptoe and crane their necks to get a good view. The doomed prisoner is led out. His crime is heresy. Across a lifetime of writing and teaching in many of Europe's greatest cities, Bruno has made many enemies. The men who lead the Inquisition can be counted among them. The specific charges they have raised against Bruno matter little. Today something strange and terrible will occur. In this ancient city shaken by the intellectual upheaval of the Renaissance and the political maelstrom of the Protestant Reformation, a man will burn for his ideas.[1]

Bruno's famous *Italian Dialogues*, two books published twenty years earlier, had helped gain him both a reputation as a freethinker and the unwanted attention of the Inquisition. In the first book Bruno proclaimed his support for the intellectually dangerous Copernican model of the solar system. Half a century earlier, the Polish astronomer Nicolaus Copernicus had placed the Sun at the center of the solar system and reduced the Earth to just another orbiting planet. This heliocentric model was considered by many to be in conflict with Scripture. It would eventually be deemed heresy. In the second book Bruno went even further, claiming a "plurality of worlds." All the stars we see at night were, he claimed, just like our Sun. Each was orbited by a family of planets, and each planet was inhabited by intelligent beings. It was a bold assertion made at a time when ideas were just as dangerous as cannons and warships. With half of Europe poised to revolt against papal authority, the Church was in no mood to entertain such freethinking. Although Bruno had not been condemned directly for his astronomical views, his bold support for contentious ideas like Copernican astronomy and the so-called plurality of worlds was a step on the path that led him to this fateful day.

Bruno is pushed to the stake, where he is stripped naked. Alongside him are a small troop of monks. Once again they ask him to recant his ideas. Gagged, he can only shake his head. The torches are lit.

On a spring day in Washington, D.C., 401 years later, the astronomer Geoff Marcy is the focal point of the National Academy of Science's 138th annual meeting. Marcy and his collaborator, Paul Butler, are accepting the prestigious Henry Draper Award for their scientific accomplishment — the discovery of planets orbiting other stars. Thanks to Marcy, Butler, and others, centuries of debate surrounding the plurality of worlds question have come to a definitive end. There *are* other worlds orbiting other stars. Marcy and Butler are hailed as heroes for their efforts as the cameras flash and the audience applauds.

. . .

The public debate between science and formal religion no longer speaks to the challenges we face as a species. The usual suspects in the conflict have been appearing onstage for decades, only the costumes changing with the times. A new perspective cannot emerge in this setting until the fog of tired definitions, outdated perspectives, and stubborn bloody-mindedness is burned away. Only then can we finally see the original and common roots of both science and spiritual endeavor. The problem is one of imagination and cultural memory.

While its formal roots were put down in the ancient world of Hellenistic Greece, science as both practice and institution came to maturity over the past six hundred years. It is true that a deeper understanding of science and religion requires reaching much farther back into history than this "modern" era, but the events of these more recent centuries are the ones that shape our expectations of their relationship. Before we can envision a new perspective we must understand how our own biases emerged from the tangle of recent historical conflicts.

We have been taught to see the debate between science and religion as a slow burn of simmering antagonism that periodically flares into the realms of open cultural warfare. Textbooks, classrooms, plays, films, and the popular media all paint the background of ideas and preconceptions about the history of science and religion. From these we form our expectations. In my own education as a scientist I was presented with countless stories about heroes and villains in the search for scientific truth. Many of the villains wore religious robes. The question we begin with is simple. How accurate is this vision? How did it emerge and who benefited from it? To gain higher ground and a better perspective we must first retrace some steps.

GIORDANO BRUNO RECONSIDERED

If Giordano Bruno were allowed to return to Earth four hundred years after his death he would be pleased with his legacy in spite of his life's fearsome end. In the centuries following his execution, Bruno's reputation has been rescued from the ignominy of his death. His redemption did not come about by a pope's edict but by advocates for free thought and science.

In the last half of the nineteenth century progressive forces in Italy embraced Bruno's story as a clear example of the Catholic Church's domination and recalcitrant hold on intellectual power.[2] Through their efforts a memorial to his memory and martyrdom was erected in the square where he was burned. In the United States during this same period Robert Ingersoll, a popular orator, politician, and self-described agnostic, praised Bruno as a champion of the intellect's search for the True. "He was the first real martyr," wrote Ingersoll, "neither frightened by perdition, nor bribed by heaven. He was the first of all the world who died for truth without expectation of reward."[3] In our own era the extent of this influence can be seen in the SETI League's Bruno award. The SETI League champions the scientific search for extraterrestrial intelligence. Each year it hands out a facsimile of Bruno's monument to worthy scientists who advance the search for intelligence on other worlds.[4] For my part, I read about Bruno many times during my education as a scientist. His tragedy was recounted in the books I read as a teenager and in my introductory astronomy class onward. He is a martyr, a fully ordained saint of free inquiry.

In truth, Bruno's story is not so simple. As much as one has to admire Bruno for the strength of his convictions and his courageous refusal to back down to power, the historically accurate story is more complex than the beatification homilies. More important, the standard Bruno martyr biography exposes the weight of cultural baggage we carry on the issues of science and religion. Within his story and the story of his legacy we can find a narrative of science and religion that is more subtle and surprising than a simple tale of warfare.

Recent scholarship shows Bruno's execution had more to do with theology than astronomy. The Inquisition's documentation of Bruno's heresy has been lost. It is difficult, then, to know exactly what offenses led to his condemnation. A reconstruction of the charges from Vatican archival documents shows that the majority of counts against Bruno concern theological juggling on issues like the nature of the human spirit and the doctrine that the earth itself has an intellectual soul.[5] These are hardly concerns that would raise a modern scientist's hackles. On top of these facts comes the inescapable conclusion that Bruno was an ass of epic proportions.

Bruno fled from one European city to another, often escaping just ahead of the latest angry benefactor. Bruno possessed a genius for driving those who helped him into dark and retributive moods. As an example consider his short stay in Geneva in 1578. After finding work as a copy editor, Bruno entered the University of Geneva hoping to find a place to teach and write. Three months later he published a mocking tract outlining twenty mistakes made in a lecture by the university's chair of philosophy. Bruno's target of ridicule was a close friend of the university's deacon, who quickly had Bruno, and his printer, arrested.[6] Within days Bruno was back on the road. Seven years later, during a stay in Paris, Bruno published a list of philosophical and theological "principles" and publicly challenged anyone to refute them. The challenge was answered by a young student who was so successful in his lecture that the audience demanded a personal response from Bruno. Bruno failed to show up. Paris soon became as dangerous for him as the other cities he had fled, and within the year he was once again an intellectual refugee.

Clearly, Bruno had difficulty making reasonable choices for himself. As the astronomer Robert Pogge puts it,

> Bruno was brilliant, contentious, and ultimately self-destructive. . . . His actions . . . reveal the very hallmark of folly, namely repeated failure to act in his own best interests even when reasonable alternatives were available. His final return to Italy (which resulted in his arrest in Venice a year later) can be seen as being motivated in part by the fact that by 1591 he had effectively burned most of his bridges behind him and thus he had little choice. In many ways, Bruno thrust himself into the flames that rose into the winter skies of the Campo di Fiore on the 17th day of February in 1600.[7]

Pogge's view passes far too light a sentence on the Church (like claiming the theft of your bike is your own fault because you forgot to lock it up). Still, his perspective on Bruno's martyrdom sheds light on our own preconceptions about the early relationship of science and religion. Though Bruno may have been a brilliant thinker whose work stands as a bridge between ancient and modern thought, his persecution cannot be seen solely in the light of a war between science and religion. That is the critical point. During these years when science was just establishing its modern practices and principles there was no well-defined war. The majority of the most fervent practitioners of science considered themselves deeply religious, and the institutions of religious power were divided in their support of their work. In the beginning it was not so much a war as a difficult but passionate marriage.

THE WORLD DOESN'T REVOLVE AROUND YOU: COPERNICUS, PTOLEMY, AND THE CHURCH

Like Giordano Bruno, Copernicus found himself at odds with the Church. There is one particular story of Copernicus and the publication of his world-shaking astronomical theory that bears on the questions I am exploring. Like Bruno's, this narrative tells us a lot about our own views of science and religion.

It took Copernicus a long time to publish his heliocentric theory. The reasons for the delay are features of the shifting way in which "conflict" appears in our view of science and religion. The story as told by my introductory astronomy professor was that Copernicus hesitated to publish his great work, *De Revolutionibus*, because he feared persecution from the Church. For me, this called to mind the vision of a fearless nonreligious scientist leading a kind of secret life beneath the thumb of the Church. Once again, however, the story is more nuanced and interesting.

Copernicus's ideas were radical. The notion that the Earth revolved around the Sun was in direct conflict with commonsense experience and a millennium of scientific and philosophical thinking. The dominant, established, and "obviously true" picture of the heavens at the time was the geocentric (Earth-centered) model of Claudius Ptolemy. Ptolemy wrote his textbook on astronomy sometime around 150 C.E. For more than 1,300 years his vision of the Sun and the planets orbiting around the Earth held sway. Throughout those long centuries, Ptolemy's geocentrism provided astronomers with the tools to calculate and predict the motion of the planets through the night sky. Those calculations proved useful enough even if they were somewhat inaccurate. It was with good reason that the Arab astronomers, who moved science forward during Europe's descent into the dark ages, called Ptolemy's book *The Almagest — The Greatest*. Although Ptolemy's theory was wrong, it was, in its way and in its time, very successful. It was both foolhardy and daring to go against Ptolemy's authority. Nicolaus Copernicus knew what he was up against.

The irony of Copernicus's story is his motivation. He did not propose his new, heliocentric model simply to make better astronomical predictions. He was striving for beauty, simplicity, and the right place to locate God. His theory that the planets, including Earth, traveled around the Sun was correct, but it was also rudimentary. His biases led him to describe the planets' orbits as perfect circles (Johannes Kepler would later discover that planetary orbits take the form of an ellipse). This mistake meant that Copernicus's model did not predict the movements of the planets any better than Ptolemy's.

The geocentric model was ungainly in Copernicus's eyes. It was not a plan that God in his wisdom would have sanctioned. Clearly, the Sun, as the source of light and warmth, was the perfect image of the divine. It, rather than the lowly Earth, had to be the center of the universe. As Copernicus wrote, "In this most beautiful temple who would place a lamp in another or better position than that from which it can light up everything at the same time? For the sun is not inappropriately called by some people the Lantern of the Universe, its Mind by others, and its Ruler by still others."[8]

Copernicus's heliocentric universe matched his sense of divine aesthetics. Others felt the same way. Honoring and glorifying God by means of the heliocentric model had the backing of many within the religious establishment, among them a number of highly ranked officials.[9] From his own writings it appears Copernicus was just as concerned about other scholars as he was about Church authorities. In the dedicatory letter to the pope at the beginning of his book, he writes of fears that his new ideas will be drowned out by the cries of others still in the thrall of the 1,300-year-old Ptolemaic system.

Thus the waters are muddied. Many Church intellectuals embraced Copernicus; others opposed his ideas to the bitter end. There can be no doubt that there were dangers from members of the religious orthodoxy, and these were already apparent for Copernicus. When *De Revolutionibus* was finally published, Andreas Osiander infamously added an unapproved preface to make it more friendly to Church doctrine.[10] But Copernicus's ideas were radical for *many* reasons. They would only grow more religiously dangerous as the decades (and the Protestant Reformation) wore on. At the time he published his book they were not, however, illegal. This would not happen until 1633, after the great astronomer Galileo Galilei had had his more famous ordeal before the Inquisition. Only then, more than half a century after its publication, would *De Revolutionibus* be added to the Church's index of forbidden books.

During the crucial time of Copernicus, Bruno, and Galileo, the war between science and religion was not something they, each deeply spiritual men, would be able to clearly recognize. What they did see was a battle over the religious worldview. Each man saw his scientific work as part of that worldview. Each man saw in his efforts an attempt to honor what he felt was the world revealed more clearly in its sacred grandeur and majesty.

Thus the image of an *eternal* and intractable war between scientific and spiritual perspectives must be seen as suspect. Most people, especially scientists like me, were raised on a steady diet of antagonism between science and religion. Recent work by historians forces us to reevaluate this long perspective. The narrative of battle we have been fed is a story of one particular religion and its institutions, namely, the Christianity of Western Europe, locked in conflict with science. The result has been entrenched attitudes that affect all discussions about science and spiritual endeavor. Reclaiming a more creative vision will, therefore, require a more nuanced telling of the tale.

Let us move forward a century or so, to the emergence of an all-out battle between science and religion during the Enlightenment. It is during this remarkable period of history that the first seeds of the warfare we recognize so easily were sown.

REVOLUTION IN THE AIR:
THE AGE OF REASON

The Enlightenment was an extraordinary philosophical movement of the eighteenth century that grew into a potent political and cultural force. This unique moment in history has been seen as many things to many people, but a useful definition is as follows: The Enlightenment was a constellation of writers, politicians, and philosophies that "rejected traditional ideas and values, emphasized the notion of human progress, and promoted the use of reason and direct observation in science."[11]

During the Enlightenment the world we recognize was given shape. It took a form that was radical in its departure from the past and traditional sources of learning. Throughout this period writers across

Europe and the New World formed a chorus of voices that rose up in
opposition to authority.[12] As Jeffrey Stout observes, "Modern Thought
was born in a crisis of authority, took shape in a flight from authority,
and aspired from the start to autonomy from all traditional influence."[13]
The importance of the Enlightenment to debates about science and
religion hinges on the fact that churches of all stamps were seen as the
authority needing to be overthrown.

The church was a symbol of the past, and the past was a fetter. "For
writers sympathetic to the revolution, the past was merely something
profoundly oppressive, wedded to ideas and values which merely per-
petuated the interests of those in power."[14] Institutionalized religion's
resistance to change, jealous protection of wealth, and entrenched
privilege made it a magnet for dissent.

Science and religion first began to be distinct cultural forces during
the Enlightenment, but we do not yet find broad discussions of "war."
That language had yet to be created because no need yet existed. As
in the fifteenth and sixteenth centuries most "scientists" of this period
still saw themselves as overtly religious and carried forward their work
to glorify the divine.

A slow but inexorable shift in attitudes began in the wake of the
successful American and French Revolutions. A new spirit of liberation
was rising. That spirit would find its hero in Prometheus, who stole
fire from Mount Olympus and gave it to humanity as a tool for its bet-
terment. As punishment for his crimes, Prometheus was chained to a
rock where each day an eagle came to feed on his liver. The myth of
Prometheus was resurrected as the gears of the industrial revolution,
driven by scientific discovery, began to catch. He was the perfect cultural icon for his time. Beethoven dedicates an entire overture to the
mythic hero. And in his play *Prometheus Unbound* the poet Percy Bysshe
Shelley captures the ideal:

> The nations thronged around, and cried aloud,
> As with one voice, Truth, Liberty, and Love![15]

Prometheus was the spirit of the new age embodied. As Alistair MacGrath writes, Prometheus "had a natural affinity with the notion of a freedom gained by the advance of science. Might not the natural sciences make available the fire necessary to liberate humanity from bondage to superstitions and irrational traditions of the past? And was not the Christian church the chief institutional embodiment of traditional beliefs and values in western culture?" What is important is that the story of Prometheus is one of conflict and war. The times were changing.

At the beginning of the nineteenth century scientific investigations were often carried out by members of the clergy. "Scientific parsons" were so common during the century that they were regarded as a well-established stereotype.[16] But during this period a profound shift in the nature of scientific activity occurred. Science became professionalized. A growing emphasis on university-based education and training shifted the gravitational center of science as an institution. As the decades progressed the model of the professional scientific investigator emerged completely distinct from the clergy. Science was in the process of establishing its authority, and the two groups were destined for struggle. Sometimes the issue came down to academic positions in the growing fields of physics, biology, chemistry, and astronomy. As MacGrath points out, "In the early nineteenth century the British Association for the Advancement of Science had many members who were clergy. . . . By the end of the century the clergy tended to be portrayed as the enemies of science and hence of social and intellectual progress."[17]

WARFARE MADE EXPLICIT

From the scientific side the explicit language of warfare between science and religion can be traced to two distinct and influential publications in the late nineteenth century. First came the 1874 publication of *The History of the Conflict between Religion and Science* by John William Draper, an English chemist. Through fifty reprintings and publication in ten languages the book solidified the vision of religion and the Catholic

Church driving the continuous oppression of science.[18] Then, in 1896, Andrew Dickson White, first president of Cornell University, published an influential two-volume work, *A History of the Warfare of Science with Theology in Christendom*. White was unrelenting in his condemnation of the Christian Church for its centuries-long attacks on the scientific enterprise. As the twentieth century opened White's work became the textbook description of the antagonism between science and religion. As David Lindberg has written, White's *History* was "treated as an authoritative source by readers who had no access to contrary opinions blessed with scholarly credentials equal to White's. [It] shaped the views of generations of educated Americans and Europeans in the twentieth century. Further defense of the warfare model was apparently unnecessary as the historic warfare of science and Christianity became an article of faith, achieving the status of invulnerability merely by virtue of endless repetition."[19]

Apparently, endless repetition works well. In the warfare between science and religion some war stories, retold as history, appear to be a kind of urban myth. When I was sixteen my grandfather gave me a copy of Bertrand Russell's *History of Western Philosophy*. Published in 1945 by the cranky but brilliant British philosopher, the book takes readers on a breathtaking journey through three millennia of human thought. Russell's tome became an instant classic and was used in classrooms as an introductory text. I poured over the book, relishing its accessible style. Russell's high regard for science and his sweeping and devastating criticisms of its opponents made those sections especially juicy for me.

In his description of Christianity's reaction to Copernicus, Russell levels his sights on the Protestant firebrand John Calvin and his famous quote, "Who will venture to place the authority of Copernicus above that of the Holy Spirit."[20] I loved this quote. Here was the fully embodied foolishness of a religion blinded, resisting the bare facts of nature. I have presented Calvin's words to my students in almost every introductory astronomy course I have ever taught. Now I find there is a small problem with Calvin's words that will force me to change my lecture notes: Calvin never said them.

While writing a book on the Copernican Revolution, the philosopher of science Thomas S. Kuhn attempted to find where or when Calvin made his infamous statement (Russell does not provide the source in his *History of Western Philosophy*). Kuhn was unable to find the quote in any primary sources. The only place he did find it was White's *History of the Warfare of Science with Theology in Christendom*.[21] After further literary sleuthing historians have concluded that the quote must be considered suspect.[22] In fact, a detailed review of Calvin's writings reveals that he never made any specific comments on Copernicus. Whatever Calvin's feelings were about the heliocentric model, this quote appears to be propaganda in a war that started long after his time.

Of course, it would be foolish to argue that the past five hundred years have been conflict-free in the domains of religion and science. I am certainly not going to act as an apologist for the religiously inspired persecution of supposed heretics (scientific or otherwise). But we are all victims of our history. The lessons of renewed historical research in the field of science and religion must be heeded. There were other forces, other social contexts, at work that would make conflict into a self-fulfilling prophesy.[23]

White and Russell show us how modern science's inherited antagonism to religion took hold. The history of the twentieth century provides a sad, ongoing narrative of the warfare metaphor made real in politics, policy, and power. This time the finger of blame points squarely to the narrow religion of a vocal but powerful minority.

THE MYTH OF PROMETHEUS:
FRUITS OF CONFLICT

> A mighty lesson we inherit:
> Thou art a symbol and a sign
> To Mortals of their fate and force;
> Like thee, Man is in part divine,
> A troubled stream from a pure source.
>
> Lord Byron, *Prometheus*

The giant eagle circles overhead in the empty sky. It is not time yet for today's horror. The mighty Titan shifts his weight and strains against the chains binding him to the living rock. Here, at the edge of the world, there is no respite. There is never any respite.

Prometheus stares across the mountains and hot dry plains beyond them. He has had time to reflect on his choices. Time passes even for a god, and in the solitary anguish of Zeus's punishment Prometheus has spent uncounted hours questioning his decision. It was not just the fire. That was simply a symbol. The fire was something concrete that humans, in their simplicity, could directly comprehend. Prometheus had not only given the poor creatures fire; he had taught them how to use it too. He had given them the tools to become more than they were. He had taught them to throw off the animal skins that stank of dried fat and blood and create something mighty, something grand for themselves. No, it was not simply the physical fire he had given them, burning torches to light their caves and cook their food. It was the fire in the mind that mattered more. Prometheus had taught humans the arts of civilization, "made them acquainted with architecture, astronomy, mathematics, the art of writing, the treatment of domestic animals, navigation, medicine, the art of prophecy, working in metal, and all the other arts."[24] He had given them the vision of what they might build when they understood how to shape the world around them. That was his real crime. That is what drove the god-king Zeus into such a rage that he cursed thunder and spat great streaks of orange fire.

Prometheus was a Titan, one of the elder gods. When Zeus rose up against his father, Chronos, in the great battle for Earth and Heaven, Prometheus joined the rebellion. Victorious Zeus had often favored Prometheus, whose name meant "forethought," seeking his wise counsel. Then Prometheus and his brother Epimetheus (afterthought) were chosen to populate the Earth with animals and give them attributes from the gods' own store of gifts. But Epimetheus mistakenly exhausted all the gods' gifts, leaving humans without benefit of speed or wings or other aids. It was then that Prometheus's loyalties shifted. Zeus had never favored humans and had even once tried to wipe them from the Earth. Zeus had ordered that humans be denied the knowledge by which they might better their lot. But Prometheus saw the creatures with compassion and looked to the future. He alone saw humankind's potential. So he stole the fire of Heaven and all the ideas that went with it and gave them to the race of men.

For this crime Zeus banished Prometheus to eternal torment, first sending him into the depths of Tartarus, the gods' own Hell, and then, later, having him chained to this mountain waiting each day for the great eagle to drop upon him and tear open his belly. Gods suffer pain just as do mortals. The raptor's claws never fail to sear as they find Prometheus's liver for its daily feast.

But in the moments just before the blinding rush of pain Prometheus's defiance rises up again and again. He is the one god who recognized what humankind could become. His gifts to the simple creatures were just; even Zeus's authority could not diminish that potential, that truth. As the eagle begins its dive toward him Prometheus burns with his own fire. "For knowledge I resisted," he reminds himself. "Till eternity, I will continue my resistance."[25]

. . .

In the early years the practitioners of science would not have recognized the notion of warfare between their work and the domain of the sacred. Their aspiration to know the world more fully was one dimension of their own spiritual (for lack of a better word) sensitivities. For them the world and its workings were a manifestation of the great power that supported all that was visible. For them the physical world was imbued with a structure that was nothing less than divine. During the Enlightenment, the practice of science became more widespread, and science became an institution in its own right. It was at that point that the story of Prometheus was remembered (as great myths always are) and given an updated narrative that fit the needs of a new age. In its struggle with religious authority science became estranged from spiritual endeavor. The story we have covered in this chapter is one of an emerging conflict as the fledgling practice of science struggled to separate itself from an aging authority. Science established its own codes, its own norms of behavior for generating truth. But while its practices were a remarkable innovation, the aspiration from which it emerged, the deeply rooted desire to draw closer to the world by understanding it, was not new. It was, instead, a continuation of an age-old imperative, a constant fire in the mind. We must now continue to follow the bright line of this fire.

CHAPTER 2

The Conflict We Know

Religion, Science, and the Modern World

I asked the bartender, what do you see?
Part man, Part monkey.
Definitely.
                    Bruce Springsteen,
            "Part Man, Part Monkey"

I like to browse in occult bookshops if for no other reason
than to refresh my commitment to science.
                    Heinz R. Pagels, *The Dreams of Reason*

Those who fight fire with fire usually end up ashes.
                    Abigail van Buren

THE ARMY OF THE SULLEN

Anthony Tucci was a hard guy. That meant he came from the Silver
Lake region of Belleville, the town in northern New Jersey where I
grew up. In Belleville you learned to stay away from Silver Lake boys
unless you wanted trouble. I wasn't looking for trouble, but my early
scientific convictions put me on a collision course with Tucci's surprisingly strong religious beliefs.

Tucci and I were in eleventh-grade study hall together, and in the absence of studying we would often fall into arguments about science and religion. In spite of his tough reputation (he and his friends had spent some part of junior high school fitting me into different-sized gym lockers) we had reasonable discussions. One day Tucci appeared in class with an air of expectation. He handed me a small booklet that he claimed proved evolution wrong. To my amazement it was a comic book. It laid out fairly specific arguments about different kinds of moths, birds, and lizards and the impossibility of their appearing via natural selection. I had to admit I did not know enough about biology to refute the claims. Then, on the last page, was a picture of an atomic nucleus and a question brimming with sarcasm. "And how can a nucleus be made up of protons if they all have the same charge," wrote the anonymous author-artist. "Even scientists know that like charges repel!" I was floored. The antievolution arguments sounded like someone had done his homework, but the potshot at physics showed me the whole thing was crap. The author had not even bothered to open up an encyclopedia. There, under "Atom," he would have found a description of the "strong force" that overwhelms the repulsive push of like electric charges and binds the nucleus together. It was the first time I had encountered such willful ignorance. I turned to Tucci and pointed out the gross stupidity of the physics in his comic book. For more than a moment the old Tucci, the dangerous one, reappeared. "Screw you," he said, and stormed off. I was lucky not to get pounded.

Anthony Tucci was a literalist in his interpretation of the Bible. It was Scripture that determined truths about the world for him, not science. In the United States it has been biblical literalists and their descendants who have propagated the worst of the warfare model between science and religion. This category includes the familiar young-Earth Creationists, who are certain that the Earth can be no more than seven thousand years old. More recently the Intelligent Design movement has relaxed the claim of literalism but not the insistence that science must be compatible with Scripture. Both the literalists and the Intelligent

Design movement represent extremes of one, traditional mode of comparing science and religion. I refer to these extremists as "the Sullen" because they harbor a kind of inborn anger toward Science. The Sullen rail against science for its stubborn refusal to bend to their scripturally based beliefs and its continuing ability to trump their best arguments. They have widened the gap between science and spiritual endeavor until it has become a chasm.

FUNDAMENTALISM, CREATIONISM, AND EVOLUTION

In 1996 a new twist emerged in the debate between science and religion with these now-famous words:

> Today . . . new knowledge has led to the recognition of the theory of evolution as more than a hypothesis. It is indeed remarkable that this theory has been progressively accepted by researchers, following a series of discoveries in various fields of knowledge. The convergence, neither sought nor fabricated, of the results of work that was conducted independently is in itself a significant argument in favor of the theory.[1]

When Pope John Paul II spoke these words in a speech to the Papal Academy of Sciences, he was giving his blessing to something the rest of the scientific community knew: evolution happened. Given the pope's remarkable statement, where is the conflict between science and religion on this most contentious issue?

In the United States the armies of the Sullen are swollen with Christian Fundamentalists. While the term *fundamentalism* has been applied to all kinds of religions, its origin stems from a set of twelve volumes titled *The Fundamentals: A Testimony to the Truth*, published in the first decades of the twentieth century. In the words of an advocate, they were written to serve "as a response to the modernism and liberal theology of the latter part of the 19th century and the beginning of the 20th. They were written in order for ministers of the gospel . . . to have at their disposal articles which would be useful in affirming . . . the fundamental truths of Christianity in the face of ever increasing attacks against it."[2]

One of fundamentalism's principal tenets is the rejection of scientific evidence for biological evolution, geologic history, and astrophysical cosmology. Like my friend Anthony Tucci, the proponents of this view, which is called Creation science or young-Earth Creationism, argue that the science that supports fields like evolutionary biology is flawed. Scientists have, for the most part, been willing to ignore the Creationists, seeing them as cranks left on the sidelines of intellectual history. Most of us thought the battle had already been fought and won in the famous Scopes monkey trial of 1925.

In March 1925 the Tennessee General Assembly published the Butler Act, making it illegal to teach "any theory that denies the story of the Divine Creation of man as taught in the Bible, and to teach instead that man has descended from a lower order of animals."[3] The American Civil Liberties Union jumped on the Butler Act. It offered to defend anyone prosecuted under the new Tennessee statute. George Rappleyea, a businessman in the small town of Dayton, Tennessee, had a keen eye for economics and saw real opportunity in the Butler Act. In a towering act of cynicism, he convinced his colleagues that the controversy of a trial would put the small town on the map. Rappleyea asked his friend John Scopes to teach Darwin's theory in the local high school, where Scopes was the football coach and a substitute teacher. Scopes agreed, and in short order Rappleyea and his beloved Dayton got all the publicity they could have wished for: the trial became an early example of a media circus.

First the nationally known populist politician and three-time presidential candidate William Jennings Bryan volunteered to argue for the prosecution. Then Clarence Darrow, a firebrand labor lawyer with his own national reputation, volunteered to join the ACLU's team representing Scopes. With these two celebrities heading the bill the trial quickly became a cultural test of "biblical values" versus science. When

Darrow, in a brilliant legal move, called Bryan to the stand as an expert on the Bible he exposed his opponent's foolishness and the inherent contradictions of biblical literalism. Scopes was found guilty, however, and fined $100. On appeal the case was thrown out by the Tennessee Supreme Court on a technicality, but the small fine would eventually be taken as a triumph for science and modernity.[4] Most important, the famous writer and wag H. L. Mencken covered the trial, and his newspaper columns condemned the Fundamentalists as intolerant, backward, and ignorant. As the decades passed advocates of the anti-evolution young-Earth movement fostered this stereotype with their own histrionics.

It would have been wonderful if the Creationists had faded into obscurity. The growth of the religious right as a conservative political force beginning in the 1980s, however, demonstrated their resilience. Creationism returned as a movement wielding no new claims to truth but something far more dangerous — political power. To combat the stereotype of backward-looking antiscience Luddites, a new form of Creationism emerged in the 1990s that would take the war between science and religion to new highs and new lows.

THE SULLEN GET SOPHISTICATED: INTELLIGENT DESIGN AND THE WEDGE STRATEGY

In the 1990s the University of California, Berkeley, law professor Phillip Johnson emerged from a personal crisis with a religious conversion and a mission. After turning to born-again Christianity, Johnson saw meaning in his life renewed through a program aimed at overthrowing "scientific materialism." This term is a Fundamentalist buzzword for evolutionary theory's refusal to include a deity in its theorizing. Recognizing that attacking science in a culture built on its fruits was bad public relations, Johnson and a cadre of conservative Christian cohorts founded the Intelligent Design movement. Backed by the Discovery Institute, a well-funded conservative Christian think tank in Seattle, Johnson and company launched a renewed attack on science that would, they thought, fight fire with fire: they would use science to show that the science of evolution was false.

The Discovery Institute established an organizational offshoot, the Center for Science and Culture (CSC), to promulgate Johnson's strategy. The CSC was intended to "overthrow scientific materialisms and its cultural legacy" by "bringing together leading scholars from the natural sciences and those from the humanities and social sciences." In a document called the "Wedge Strategy" they outlined their five-year plan, which included scientific research, conferences, publicity, and, finally, planning for the court cases to come.[5] In Johnson's words, "We call our strategy the wedge. A log is a seemingly solid object, but a wedge can eventually split it by penetrating a crack and gradually widening the split. In this case ideology of scientific materialism is the apparently solid object."[6]

For the most part the group was successful, and the use of the term *Intelligent Design* in the public debate is a testimony to its reach and backing. There was one notable exception to its success. The CSC was never able to produce a single real scientific paper. Intelligent Design could not gain traction within the scientific community because there was no science in it. The arguments over design, the arguments "for" design, and the arguments "from" design have been around for a long time. In spite of the CSC's insistence, nothing in those arguments could overcome the overwhelming evidence for a biology based in evolution.

EVOLUTION AND DESIGN

Anyone who takes an introductory philosophy course in college covers arguments for the existence of God. This is an old debate, dating back at least a thousand years. Among the more famous "proofs" for God is the so-called Argument from Design. It has taken many forms over the years, but it was most succinctly and brilliantly argued by the English theologian William Paley in 1802. It was Paley who came up with the famous watchmaker analogy used so often against evolution. The argument, simply put, runs like this.

Imagine you are walking through the countryside and happen upon a pocket watch lying in the grass. It would only be natural, Paley argues, to assume that something so intricate had been made by the hand of an Intelligent Designer. Even if we do not know what a pocket watch is for, we could still infer that, unlike the rocks lying around it, the pocket watch is full of interconnected moving parts and therefore must have had some design in its making. As Paley writes, "The watch must have had a maker: there must have existed at some time and at some place or other, an artificer or artificers . . . who comprehended its construction and designed its use."[7] Paley then draws the analogy between the watch and biological structures such as the eye. He concludes that they too were designed by an intelligent creator.

By all accounts Paley's form of the Argument from Design was masterful. Even Richard Dawkins, the English biologist called by some "Darwin's Pit Bull," gives Paley high praise. "[He] had the proper reverence for the complexity of the living world and he saw that it demands a special kind of explanation," Dawkins writes.[8] Paley's argument held sway as an explanation for biological forms until Darwin presented his alternative.

In 1859 Charles Darwin published his *Origin of the Species*. Like Copernicus's *De Revolutionibus* and Newton's *Principia*, Darwin's book stands as a monument, a turning point, in human thinking. For Darwin natural selection and not purposeful design led to the diversity of biological forms. In this process, free from intervention by God, small changes in an organism's structure and function become favored by changing environmental conditions. For example, the eye took its form through time because each small change made the organism and its heirs more likely to survive. Changes that did not work this way led to death and extinction.

Darwin's explanation became wedded to the growing recognition in Victorian science that the Earth itself had existed for a long, long time.

Geologists in the late 1800s followed multiple lines of evidence to the conclusion that Earth had existed for millions of years at least, rather than the seven thousand years calculated from Scripture.[9] Allowing natural selection to work on vast, nearly unimaginable time scales pulled so many pieces of evidence together, from disciplines ranging from paleontology to physiology, that biologists quickly came to a consensus about the importance of Darwin's insight. Over the next century and a half advances in the study of life's microscopic structure and function contributed significantly to the understanding of evolutionary processes. The discovery of DNA as the molecular blueprint for living systems expanded the evolutionary paradigm. The ongoing "genetic revolution" added the insight that variations in organisms occur through "changes, or mutations . . . in the nucleotide sequence of DNA, the molecule that genes are made from."[10] These changes in DNA can now be detected and described with great precision using modern technology. Evolution is a theory no different from aerodynamics or electromagnetism. It is a coherent, consistent body of knowledge that forms the bedrock on which all of biological science rests.

DESIGN VS. EVOLUTION

In spite of the centrality of evolution in biology, the Center for Science and Culture managed to accumulate a list of scientists with credentials in a variety of fields. Through them, it challenged the foundational principles of evolutionary science. Using "real" scientists (whose religious convictions underpin their perspective) has proven an effective public relations strategy for the Intelligent Design movement. It provides the movement with the air of respectability that the Creationist movement lacked.

Michael Behe, for example, is considered a star among the stable of CSC scientists. Behe is a professor of biological science at Lehigh University and the author of *Darwin's Black Box: The Biochemical Challenge to Evolution*, an Intelligent Design best-seller. Much of Behe's anti-Darwin argument is based on a concept he calls "irreducible complexity." On this view, biological systems are simply too complicated to emerge from successive adaptive changes. Jonathan Wells, another scientist used frequently by the CSC, holds Ph.D.'s in biochemistry and theology and offers his own detailed arguments against evolution. He has written widely on his view that natural selection cannot lead to new species. Both Behe and Wells hold that changes within species through natural processes may be possible but that the creation of new life-forms requires some form of Intelligent Design and hence an Intelligent Designer.

The arguments put forth by the scientific members of the CSC demonstrate their scientific sophistication and their understanding of the nature of the game. They use the language of evolutionary theory and refer to its data. But this sophistication does not change the unfortunate fact (for them) that their science is fundamentally flawed. It is one thing to understand the scientific basis for a particular argument; it is quite another to come up with a creative and compelling hypothesis that stands up to existing or new data. Science is not simply an exercise in rhetoric, as much as Phillip Johnson, with his legal background, might hope it is.

It turns out that there is, fundamentally, no fundamental evidence for Behe's irreducible complexity. This is a point that has been worked over by numerous scholars.[11] Wells's arguments that natural selection works only within species is also demonstrably at odds with the vast weight of scientific data from diverse fields.[12] On both counts Behe and Wells reveal the real problem with researchers associated with the CSC. They produce no new science. They refer to data, but their ideas do not do a better job of accounting for it. The currency of scientific life, the way in which we scientists measure the success of our ideas and careers, is the refereed articles we publish in reputable journals. A search of the literature carried out by Barbara Forrest, a professor of philosophy at Southern Louisiana University, shows that as of 2001 none of the scientists associated with the CSC had published any peer-reviewed articles on their alternatives to Darwin. As Gertrude Stein once quipped, "There is no there there."[13]

Meanwhile the wedge strategy has been successful on the political front, where scientific truth need not enter into the argument. In 1996 the Kansas Board of Education passed new standards forcing Intelligent Design into the classroom alongside evolution. Similar measures have been proposed throughout the United States. In a clear measure of the political nature of the battle, President George W. Bush expressed his support for teaching intelligent design. Happily, in some cases reason prevailed over political bullying. In a closely watched court case in 2005, a U.S. District Court ruled on the case of *Kitzmiller v. Dover Area School District*. The court stated that the Pennsylvania school board's policy promoting Intelligent Design violated the state constitution and that Intelligent Design is not science. The ruling barred the teaching of Intelligent Design in public school science classrooms. That, however, was just a single victory in what promises to be a long war.

One of the hallmarks of scientific practice is its extraordinary capability for self-correction. If one idea fits the data better than another, eventually that theory will get its hearing. Scientists make careers on shooting holes in old ideas and building new ones. If I could find a better description of space, time, and gravity than Einstein's I would be happy to go against the big guys in spite of the obvious resistance. If there were phenomena, meaning real data, that would allow me to refute Einstein I would certainly have a case. In the realm of Intelligent Design there are no "better ideas." The paradigm of evolution by means of natural selection is more effective than anything the Intelligent Design scientists have come up with in spite of their well-funded claims to the contrary. Worse, even a nonbiologist can see the weight of religious bias in their arguments.

Beyond specific results, there is the brutal philosophy of warfare inherent in the Intelligent Design movement. This bears importantly on our own understanding of science and religion. In the minds of its adherents, Intelligent Design is part of a holy battle for Truth, and all opposition must be crushed. The political will and reach of the Intelligent Design movement demonstrates the vast gulf that now characterizes popular discussion of science and religion. The sheer volume these voices bring to the marketplace of ideas all but ensures that more reasoned, sensitive, and nuanced views on science and religion will be drowned out. If we are to create a different perspective on science and spiritual endeavor, then it will, unfortunately, have to be forged in the midst of these battle-hardened combatants and their constant preparations for war.

The armies of the Sullen maintain and support the language of warfare. They represent a dangerous challenge to the very core of the scientific enterprise. Not all popular perceptions of science and religion take such a combative view, however. At the same time that the Creationist movement found renewed energy in its attack on science, an entirely different perspective was gaining ground on the other side of the philosophical-political spectrum. The so-called New Age movement has its own take on science and spiritual endeavor, which has been, in general, just as wrong as that of the Creationists. We will need to understand a bit of its perspective also if we are to forge a more thoughtful and vital perspective on science and spiritual endeavor.

A NATION OF SILLINESS: EASTERN RELIGION, NEW AGE PHILOSOPHY, AND THE QUANTUM PROSPECT

In the early 1980s two books, *The Tao of Physics* by Fritjov Capra and *The Dancing Wu Li Masters* by Gary Zukov, permanently changed the popular conception of the intersection of science and spirituality. These books, and others that followed, sparked widespread interest in the confluence of science with Eastern religious worldviews. Quantum mechanics, the physics of subatomic phenomena, took center stage, with its strange wave-particle dualities, uncertainty principles, and collapsing wave functions. Now terms such as *quantum healing* and *observer phenomena* have entered the lexicon of popular consciousness in certain circles. Like the Judeo-Christian-centric view, this line of reasoning spans the gamut, from extremely serious discourse to just plain silliness.[14] And like the young-Earth and intelligent design debates, some of its claims are so grossly wrong that practicing physicists like me can only shake our heads in despair.[15]

The influence of this perspective, which I refer to as "the Silly," can be seen in a variety of ways. Look at the catalog of an adult learning center or a large retreat center, for example, and you will likely find classes linking quantum physics with everything from past-life regression to crystal energy channeling. In 2001 the film *What the Bleep Do We Know* became an underground success, making the rounds of theaters across the country with a retinue of experts who made the leap from quantum physics to positive spiritual evolution. It is remarkable how widely the buzzwords of "quantum spirituality" have diffused. I once had the distinctly unreal experience of being told that I should purchase a magnetic bed (a bed with magnets glued to the frame) because quantum physics proved it would increase my well-being.

Attempts to link Buddhism, Yoga, or generic New Age philosophies with scientific paradigms differ from the angry inward focus of the Fundamentalists and Intelligent Design crusaders. The Sullen reject a dominant paradigm (evolution) because it is at odds with their religious perspective. The Silly embrace a dominant paradigm (quantum mechanics) as confirmation of a cherished spiritual worldview. The problem is that in spite of cheerful aspirations to a brighter day when "we are all connected," linking quantum physics with specific religious doctrine is fundamentally flawed. In works by writers such as Deepak Chopra we find quantum mechanics used to prop up ideas that have no connection to it. These attempts to force quantum physics into the strictures of a great spiritual tradition such as Buddhism or more recently imagined New Age philosophies miss the essential point that in science (as in true spiritual practice) the world cannot be made to fit a preconceived ideology. To understand the problem more clearly it is worthwhile to take a short detour into the realm of quantum mechanics. The importance of at least touching on quantum physics is summed up in a famous quote attributed to Richard Feynman: "There are four or five people in the world who understand Einstein's theory of relativity, but nobody understands quantum mechanics." Niels Bohr, one of the early founders of the field, also spoke of the unnerving perspective of the new physics: "Those who are not shocked when they first come across quantum theory cannot possibly have understood it."[16] The behavior of the universe in the domain of the very small is wonderfully strange indeed. It is not, however, the basis for a new religion or the confirmation of old ones.

At the end of the nineteenth century physicists began constructing a new generation of experimental devices that allowed them to explore the world on increasing smaller scales. As their experiments reached down to regimes measured in the billionths of a meter they were able to probe a staggering array of new phenomena: the nature of the atom, the way matter emits light, the physics of solids at ultra-low temperatures. These were domains of nature to which scientists had not previously had access. The experiments unveiled bewildering new behaviors that proved exceedingly difficult to understand. As physicists confronted the world through these new instruments they were forced to radically alter their approach to, and conception of, physical reality.

Attempts to make sense of the experiments using the physics of the day, what we now call classical physics, failed entirely. In desperation, and through an amazing display of human creativity, scientists created an entirely new branch of physics, quantum mechanics. The difference between classical and quantum physics can be seen as the difference between commonsense notions of how the world behaves and something else entirely. Our common sense and our classical physics derive from a world of things that are "about our size" (for a physicist this can mean anything from millimeters to mountain-sized objects). As kids we play with rubber balls, ride our bikes, and scream our heads off on roller coasters. All of these experiences condition us to expect certain kinds of behavior from the world of "things."

A classical physicist might imagine the atomic world made up of particles (electrons, protons, etc.) that look and behave much like microscopic billiard balls. These tiny spheres bounce into each other, spin, hold an electron charge, and react to gravity or magnetic fields. Such tiny specks of matter would be thought to exist "out there," independent of anyone performing experiments. They would have definite properties, and those properties should be measurable to any degree of accuracy you might demand. The problem with this kind of "common sense" was that it did not hold up on the microscopic level. Physicists quickly found that it was impossible to build working, predictive theories — that is, mathematical models — using these little spheres. Nature, it seemed, was not built that way. As Bohr once said, "Atoms are not things."

To accommodate the new data physicists were forced to abandon the billiard ball picture of reality. What replaced it was not, however, a new set of pictures. The language of quantum mechanics that emerged was rooted in an abstract mathematical formalism that did not allow you to imagine an electron the way you imagined an ordinary object like a chair or a rock. While the mathematics borrowed broad ideas such as the conservation of energy from classical physics, it did not provide a way to "picture" what was happening the way classical physics did. You could not imagine, or sketch on paper, the "things" quantum physics described. The new rules that physicists were discovering in their data appeared within a theoretical framework that absolutely restricted what they could say about the stuff they were studying. One example of this development is the Heisenberg uncertainty principle. In constructing a theoretical account of atomic phenomena, Heisenberg found that his description forced particles to have an inherent fuzziness. Certain pairs of properties, like position and momentum (related to a particle's speed), could not simultaneously be known exactly. This was weird. From a classical physics perspective the position and speed of a cue ball rolling across the pool table can, in principle, be known at any instant to whatever accuracy you care to measure. After all, doesn't the cue ball *really* have properties like position and speed? At any moment, doesn't the cue ball exist at an exact location and move with an exact velocity?

Heisenberg discovered that this kind of question does not work on the level of atoms. Subatomic things do not have properties the way macroscopic objects appear to have. Heisenberg's uncertainty principle does not say that there is a limit to our measuring device; instead it says that there is a limit to measurement itself. It is as if nature had set bounds on the idea of what an electron really looked like. In a sense electrons don't look like any "thing." As a professor of mine once said, "In quantum mechanics an electron is simply that to which we ascribe the properties of an electron." This is the essence of Bohr's comment above. As the quantum mechanical description of nature matured it became harder to think of an electron as a "thing" in the classical sense of a table, chair, or cue ball.

The world of quantum mechanics is strange to everyone when first encountered. Two aspects are especially weird and often attract the attention of people interested in spiritual possibilities. The "observer effect" is a statement about the nature of quantum systems before a measurement is made. In quantum physics particles are interpreted to exist only as probabilities until a measurement is made. An individual electron can be thought of as smeared throughout space until an instrument detects it and forces it from the state of the potential into the state of the actual. Thus it appears that the observer, the one making the measurement, has disturbed and altered reality. Along with the observer effect there is the paradoxical wave-particle duality. Experiments can show electrons to behave like particles — little chunks of matter that occupy only one location at one moment in time. Experiments can also be done that show electrons behaving like waves — spreading out from a point and existing in many places at once. The behavior you get seems to depend on the experiment you choose to perform. Once again it appears that the observer affects the observed. One could go on and on in this vein. There are many other "paradoxes," many ways in which quantum mechanics presents us with a world that seems radically unlike our everyday experience. What matters for our discussion of science and spirituality is what we do with this weirdness. What do we think we learn from it especially as it pertains to the domain of science and religion?

The apparent departure of quantum mechanics from the classical worldview and its external, objective reality sitting "out there" has fostered much philosophical and religious theorizing. In particular, the New Age movement, which emerged from the 1960s and 1970s, has embraced the apparent weirdness of quantum physics as proof that consciousness is more important than matter and that the world is imbued with spiritual realities of great and grand potential. It seems very exciting. Unfortunately, it misses the essential point that quantum mechanics doesn't really say anything.

In truth, quantum physics is "only" a powerful calculus for describing experiments—meters and dials and digital readouts. Its weirdness is what makes it so much fun. There is nothing, however, in that calculus that confirms New Age, Hindu, or Buddhist worldviews. Quantum physics raises vital philosophical questions that do touch on some of the deepest questions we can ask, but it does not answer those questions. There is the mathematical calculus inherent in quantum physics, and then there are the interpretations of that math. The point too often overlooked is simply this: the history of quantum mechanics is littered with interpretations.

It is true that some interpretations of quantum physics give consciousness a central role in fixing reality. But there are others that are quite mundane and attempt to preserve a classical worldview with no role at all for the observer.[17] Still other interpretations posit radical ideas such as the existence of infinite parallel universes. What matters for the religion and science debate is that these interpretations must be added to the mathematical theory like kids playing pin the tail on the donkey. In spite of claims made within some of these interpretive frameworks, there is, as yet, nothing inherent in quantum calculations about consciousness.

The film *What the Bleep Do We Know* typified the New Age enthusiasm for quantum physics. As one "expert" interviewee stated, "Every morning I get up and alter the quantum fields around me." Watching this, I had to work hard to keep from hurling my jumbo popcorn at the screen. This guy has no more control of the quantum fields around him than I have over the gravitational field around me. Maybe there is some field he controls, but it is not anything you can measure. It is simply wishful thinking substituting for an honest assessment of real parallels between science and human spiritual endeavor, and that is where the convictions of the Silly go wrong. In science and spiritual practice you cannot substitute hope for hard work. Worse, too much of the discussion of quantum mechanics and Eastern religion relies solely on the latest results of science to shore up spiritual worldviews that should not require them. With this emphasis on results the Silly share much in common with the Sullen, even if they would not recognize it. Building a fundamentally different approach to parallels between science and spiritual endeavor will require going beyond results to something more sustaining and substantive.

BEYOND WARFARE, WEIRDNESS, AND RESULTS

> The birds have vanished into the sky
> Until the last cloud drains away.
> We sit together, the mountain and me,
> Until only the mountain remains.
>
> Li Po, quoted in Stephen Mitchell,
> *The Enlightenment Heart*

Popular discussions of science and religion almost always assume that what matters are results. It is what science "says," its published results, that needs to be compared with religion. This emphasis on results chains debate and discussion to a tired standard that is no longer relevant to the needs of a global culture facing global issues of survival.

For the past four hundred years, the traditional, or classic, discussion about science and religion has been an argument between the monotheistic God of Abrahamic tradition and the ever-expanding power and vision of science. The Judeo-Christian God has been the center around which the debate orbits.[18] Thus the debate about science and religion has traditionally meant priests, pastors, and philosophers of Judeo-Christian heritage endlessly debating if, how, and when the laws of natural science limit God's powers or vice versa. Early on, Eastern traditions such as Hinduism and Buddhism were not even acknowledged as religions as such and thus were not considered throughout much of the history of the debate. The entrance of Eastern religions into the discussion is a recent phenomenon.

Almost always the traditional debate in science and religion has been religion reacting to the latest scientific research. Should the Catholic Church accept Copernicus's new conclusions about the solar system, or should it brand its adherents heretics? Should Anglican pastors preach acceptance of natural selection after the publication of Darwin's works, or should they fight for divine design? Always the battle has raged over what the science of the age says, or does not say, in relation to claims interpreted from Scripture. Always it is only results that matter.

Results are the problem for young-Earth Fundamentalists and the Intelligent Design movement. They are interested in what science says, not its deeper connections with the source of people's aspiration to the True and the Real. If tomorrow's scientific journals gave proof of Scripture or of God the antagonism would end. This focus on scientific claims marks the failure of traditional popular modes of comparing science and religion. Although the extremes of Creationism tend to give everyone interested in issues of God and science a bad name, there has been a long tradition of highly sophisticated discussions of science and Abrahamic religion. Contemporary scholars like Ian Barbour have provided insightful distinctions regarding how science and religion need to be compared. His work and that of others such as Alister McGrath are thoughtful encounters with the two perspectives and stand in stark contrast to the bullying and bias of the Intelligent Design movement.

Still the weight of scientific results and theological imperatives burdens the debate.

The more recent discussions of correspondence between quantum mechanics and Buddhism or Hindu mysticism also focus on results. As with the traditional science and religion debate what matters for enthusiasts of "quantum spirituality" is the apparent correspondence of atomic physics with Eastern worldviews. Once again the emphasis on results is misguided. What would happen, if one of the more mundane, prosaic interpretations of quantum physics were proven true? Would Yogis abandon their ashrams and Zen Buddhists abandon their zendos, each to practice something more "realistic"? Would New Age parishioners give up their magnetic beds and settle back into the staid religions of their parents? Not likely.

The fact is that Eastern religions do not need quantum physics any more than quantum physics needs Eastern religions. Many people come to their profound sense of life's sacred character from a route that has nothing to do with facts about physical reality. As individuals, they yearn for an understanding of their deepest sources of inspiration and a connection with what they apprehend as sacred. I include myself in this category. We are, all of us, immersed in a culture that apprehends much of our collective experience through the gateway of science, yet that gateway seems closed to a simultaneous apprehension of the world's spiritual dimension. I believe the sources of aspiration can be apprehended in both science and spiritual endeavor. Recognizing its form and, ultimately, its power to reshape action will require a different kind of effort than what has gone before.

THE MYTH OF PANDORA:
TRAGEDY OF THE LITERAL

The women were gathered around the large clay urn. Their discontent had grown with each passing day, and now they were close to desperation. It had been weeks since Epimetheus, brother of wise Prometheus, arrived with his beautiful companion and her mysterious earthen jar. Since that time so much had changed — and not for the better, not like when Prometheus had come with his great gifts. The wise god had brought fire and taught men and women the arts that had lifted many burdens from their backs. But now a foul discord lived in the women's hearts. It had entered unbidden and grew rapidly like rank weeds. The women could not understand its source, but they were sure that Pandora's large unadorned urn held the answers to their despair.

Whereas Prometheus was farsighted his brother was slow of wit and forgetful. After his brother's punishment Epimetheus had withdrawn into a valley, living among the ruins of a temple long forsaken by the Titans. Zeus was all too aware that Epimetheus was simpleminded. He knew Epimetheus would quickly forget his brother's last grave warning, "Beware the gifts of Zeus." Still raging from Prometheus's betrayal, Zeus set in motion a plan to fight fire with fire. Prometheus had brought men a gift they prized. Now Zeus would use Epimetheus to bring humans his gift, one they would regret forever.

Pandora was the work of the gods. She had been fashioned from clay by the lame forge god Hephaestus. Aphrodite, goddess of love, laid charms on the new maiden's lips and the Graces fashioned necklaces and a golden crown to adorn her head. The beauty of Pandora was great, enough to inspire reverence in any mortal whose gaze fell upon her. Then Zeus commanded Hephaestus a create a clay urn for Pandora, and into it Zeus poured his special gifts to humankind.

Pandora and her vessel were brought to Epimetheus as a gift of reconciliation from Zeus. The forgetful god was instantly taken with the girl. Pandora was herself a simple creature, content to marvel at the world and its beauty. Epimetheus delighted in Pandora's company and spent many days with her in his valley. In time Epimetheus thought to bring Pandora to the homes of humans so that they too could appreciate Zeus's great gift. He brought the clay urn along as well, for all the things of the golden maiden were precious.

At first the men and woman looked upon Pandora with delight.

Her garments were of a kind they had never seen, her golden crown so luminous, the beauty of her lips and skin so unlike their own rough features. For many days and nights they were happy simply to be in her presence. But in time the men forgot their work. All their speech turned to Pandora and her glories. They forgot their fields, the tasks at hand in the forge, the houses needing building or repair. The women saw this. They wept, grew angry, or became sullen at the mention of Pandora's name.

The women gathered again and again. "We are in great need," they said to each other. "We are without hope," they wailed. They felt Pandora's beauty as a great thorn in the flesh of their lives and cried out for a single answer that could salve their pain. At last a woman who was considered wise said, "The clay urn that Pandora brought with her. Surely it holds the secret to her grace. Let us go to it and open it. There we can find our salvation." Together they found the place at the edge of the village where Epimetheus had left the clay vessel, which stood as high as their shoulders. As they gathered before it each woman gazed with longing. During the trip from the valley the lid on the vessel had come loose. That was how the first of Zeus's malicious gifts, Self-Thought, had already escaped and fashioned itself to the women's hearts. Then, determined to find salvation, the women set upon the clay urn and cast it down. But instead of finding salves, charms, and ointments the urn held things black, gray, and red. These were the Cares and Troubles. They were Zeus's dire boon for the race of man. The dark things flew and crawled from the urn. They attached themselves to the women and quickly found the men. Sickness, War, and Strife between friends were let loose to create their havoc.

The homes of men soon darkened with these Cares and Troubles. Realizing his folly and his failure to heed his brother's wise advice, Epimetheus stumbled from the village. In his grief he lost his footing and fell from a high cliff into the sea. Despair threatened to engulf the world of men, but Zeus had included one other gift in Pandora's clay vessel, and, caught under the lip of the urn, it was the last to crawl out

blinking in the hard light of midday. Zeus's final boon to humanity was
Hope. When a weeping woman found this beautiful living thing she
brought it to the rest of her race and they rejoiced. Now, at least, they
could see an end to their troubles. Hope, the gift that had been caught
under the rim of Pandora's clay urn, stayed with men and women long
after the simple Pandora returned alone to the valley. Hope stayed with
humanity as a salve to their afflictions. It lived with them behind the
threshold of their homes, serving as a guide to their best instincts and
pointing the way to what they still might become.

. . .

The fixation on results in the debate between science and religion has
brought its own set of Cares and Trouble. The army of the Sullen has
been vengeful in its singleminded and erroneous crusade. The nation
of the Silly has been all too willing to abandon critical thinking in
their desire to see cherished belief validated. And the institutions of
science itself have unwittingly developed a practiced scorn for anyone
who seeks to broach the questions. Now the gulf between science and
spiritual endeavor is so vast that any attempt to raise the issue, however
thoughtful it may be, is met with suspicion, rancor, and uncritical bias.
The situation is, in a word, afflicted.

Might we find a way to bridge this chasm, to reach a more meaningful understanding of the relationship between science and spiritual
endeavor? Such is the function of Hope, the last inhabitant of Pandora's
"box." A new perspective *is* possible — if we are willing to venture
beyond the emphasis on results. Our vision of ourselves within science
and spiritual endeavor *can* be broadened if we are willing to look to the
very experiences that drive the constant fire, our immutable desire to
know and to understand. In a global culture we must search for what
is common to humanity. We can begin our search by turning in a new
direction and bringing the content and character of "religious experience" into focus.

CHAPTER 3

Science and the Sacred

Telescopes, Microscopes, and Hierophanies

The sacred is equivalent to a power and in the last analysis to reality. The sacred is saturated with Being.
Mircea Eliade, *The Sacred and the Profane*

The true mystery of the world is the visible, not the invisible.
Oscar Wilde, *The Picture of Dorian Grey*

For an instant I thought of fire; the next, I knew that the fire was within myself.
R. M. Bucke, quoted in William James,
The Varieties of Religious Experience

It's the fourth day of a weeklong conference on star formation. More than one hundred fifty astronomers from around the world are packed into a lecture hall on the campus of the University of California, Santa Cruz. Outside the day is blinding in its perfection. If you face west you can see the Pacific Ocean. Face east, and you stare into a redwood forest. The sunlight is sharp in the cloudless sky. It would all be wonderful if any of us actually went outside to admire it.

While the day is warm and bright and graced with a mild pacific breeze, we are hunkered down in a stuffy room with the lights turned off. For the past four days we have all but lived in the dark lecture hall as one PowerPoint slide after another clicks by. The paradox of the beautiful day outside and the darkened room inside is not lost on us. It has been a long week, and our brains are entering the oatmeal phase particular to conferences. We have sat through a multitude of talks with a multitude of equations and graphs with squiggly lines superposed on squiggly lines. Now it's 2:30 in the afternoon, and you can feel the mental exhaustion settling in. We're losing our focus. The current speaker finishes his presentation and a new one comes to the podium. After a moment of fiddling with the computer projector, an image appears on the large screen overhead. A collective and barely perceptible gasp crosses the room. Suddenly we are awake again.

The Hubble Space Telescope image of a vast interstellar cloud hangs before us in bright hues of cobalt, scarlet, and jade. The cloud appears scalloped. Concentric rims of bright gas form terraces and towers that radiate from the center. A bright young star dominates the image and the cloud itself. Just a few hundred thousand years ago this massive sun was born from within the cloud. Now powerful winds and torrents of radiation stream off the young star. Gusts of plasma and ultraviolet radiation reach across light-years, tearing the cloud apart, sculpting it into the variegated chaos we see before us. The image is stunning in its beauty and implied power. With this vision of cosmic birth and destruction we all, collectively, catch our breath. In an instant we are returned to the inspiration that brought us all to this job in the first place. In the face of this image we are all kids again, standing under the night sky as it leans down hard on us and asks, "What am I?"

· · ·

What happened to my colleagues and me in the instant the Hubble image flashed on the screen? Was it simply a moment of intellectual excitement? Was it our reason excited by the possibility of new scientific insights and new intellectual conquests? Was it the excitement of our passion for astronomy and a rapid renewal of our lifelong interest in all things celestial? Was the experience related only to the life of the mind and the process of scientific activity?

Scientists are not the only ones who catch their collective breath before these pictures. Hubble Space Telescope images have found their way into thousands of magazines, TV documentaries, and even grunge rock CD covers. There is something unique and powerful about these pictures, with their astonishing scale and detail. I have seen and felt moments like the one at the Santa Cruz star formation conference repeated many times among nonscientists when I give popular talks. The momentary hush and the gasp that follows are involuntary. There is an experience that is drawn out of these images, these fruits of science, that many people in the audience recognize as "spiritual" even if scientists would not call it such. I know this because they tell me so afterward. "It made me feel so small and yet part of something so much bigger," a woman holding the hand of her five-year-old once said to me after a talk I gave on extrasolar planets. On another occasion, after a presentation on cosmology, a large, older man with a gruff manner approached me and said, "I have to tell you, I'm not religious, but some of those images make you feel like you're staring at God."

The world can reveal itself to us in many ways. Some we recognize as lying within the domain of spiritual experience; some, outside that domain. Science is seen, especially by scientists, as lying in the latter camp. This is the root of the problem when it comes to renewing perspectives on science and spiritual endeavor. We are born into a culture where the longing for spiritual understanding stands in opposition to the search for truths of the physical world. When intelligent and sensitive people speak of their experience of the world's sacred character their language can only be directed toward the interior domains of spiritual life. There is no language for seeing their experience of the physical world, the domains of science, as part of a lived sense of their own spiritual dimensions. There is no language that allows people to see what happens in science as part of the continuum of human experience that makes life and Being sacred to us. This, to my mind, is a tragedy and a failure of both our imaginations and our memory. There is, I believe, another way to understand science and its relation to spiritual endeavor, a way that can draw them into an active, vibrant parallel without forcing jealous comparisons of their claims or results.

We have already seen that the history of science and religion is more complex than simple stories of warfare. In spite of dangers from various camps in the Catholic (and Protestant) Church many Renaissance scientists had their own deeply spiritual or religious reasons for pursuing their investigations. The language of warfare between science and religion became fully formed only during the early part of the twentieth century. Now it is the only language we expect. We have also seen that the battle lines are almost always drawn around results — what science says versus what religion says. Rarely do we see public discussion about other ways of taking on the issue. How about asking what religion calls to in us as individuals? Or better yet, what people experience when they have "religious" experiences? The emphasis on comparisons of Scriptures and doctrine on one side and scientific results on the other relentlessly narrows the discussion. There are other ways of looking at science, and there are other ways of looking at religion.

The first thing most people think about when they hear the term *science and religion* is a battle between science and the Judeo-Christian God. The God they imagine tends to be willful. "He" is the one people pray to and the one who intervenes on their behalf. Starting the conversation from this point poses a real problem for many scientists, and I count myself among them. We scientists find it difficult to reconcile the absolute and timeless laws of science with a personal God who seems to step into and out of the picture. There is also an obvious problem with this take on science and religion: specific ideas about God vary among specific religions. The conception of Deity in Hinduism is vastly different from that held by Jews, Christians, and Muslims. And of course Buddhists do not even have a conception of God as such. *Religion* in the science and religion debate cannot just be about God because religions do not all think about him/her/it in the same way and some do not think about him/her/it at all.

Let me be clear on this point. Obviously, God is central to many people's thinking about religion. But the content of religion is not only about concepts of the Deity. What I want to explore is an aspect of religion that cuts across human culture and millennia of cultural evolution. It does not diminish belief to see that the imperatives of religious feeling have other facets. In that spirit can we speak of religion and, more to the point, of science and religion without a focus on the concept of and a belief in God? The answer is yes, absolutely. Some of the most vibrant thinking about religion even within the Judeo-Christian tradition has done just this. What matters to these thinkers are not academic ideas about God but something more primal and original: religious experience. For the past two centuries there has been a bright line of scholars, William James, Rudolf Otto, Mircea Eliade, and others, who have seen religion and the activity of spiritual life through the focal point of experience. For them what matters is the elemental fact that people have deep and abiding experiences of wonder, mystery, and awe. These cannot be recognized as anything other than spiritual or religious. In these experiences the world's sacred character is encountered, and people feel they have come to know something about the world.

A focus on experience marks a very different approach to understanding religion from what most people, most scientists, are used to and from that which underpins years of public debate on science and religion. If we wish to find a different vantage point to renew our perspective on science and religion, then the focus on experience can be a first step, a trailhead that will lead us on a new, deeper, and more fecund path.

So let us step off the old road and set off in a different direction. Let us recognize that our aspiration to understand the world expresses itself in both abstract thought and direct response to experience and that they cannot be easily separated. Our capacity to respond to the world's voice through spiritual or scientific language is rooted in the constant fire of aspiration. Developing a deeper understanding of its call, and our response, is the connection we are searching for. Let us begin by learning the language of key scholars in the study of religion. Through their works we will find a vocabulary to speak of science and spiritual endeavor that honors the unique capacities of each while embracing a wellspring of deeper, older connections.

PROFESSOR JAMES GOES TO SCOTLAND

No one can emerge from a consideration of religion without thanking William James.
Ursula Goodenough, *The Sacred Depth of Nature*

Each day the audience seemed to get larger. Outside the lecture room, men in bowler hats and ladies in wide dresses enjoyed the warm Edinburgh spring of 1902. As the carriages trundled down streets alive with new green leaves and fragrant with first blossoms, William James sat at the front of a packed lecture hall working his way through his Gifford Lectures. The famous American psychologist had been invited to Scotland to present his views on religion. Now as the days drew on and Edinburgh woke from its winter gray, James guided his audience through terrain they had not seen before. A new perspective on the roots of religion was being developed before their eyes. The world would not soon forget what happened in this room. But the mild spring outside belied James's struggle to create these lectures. He had spent a year wandering across Europe from one spa to another trying to ease his ill health and complete the work. He knew he was pushing himself too hard, but he also knew the lectures were worth the effort. They were an honor and an opportunity he could not miss. Since boyhood he had thought long and hard on this subject. Now he had the chance to put those thoughts together into a coherent whole. The Gifford Lectures were William James's last chance to gain some understanding on the great subject for himself.[1]

William James, known as the "adorable genius," was born to a life of inquiry. He was the first son of a wealthy, eccentric New York City family. His father moved easily in New York society and was a passionate amateur theologian. There was a constant stream of guests at the James house, including many of the nation's intellectual elite. The heady atmosphere had its effect on all the James children. William's brother, Henry, went on to become one of America's most famous novelists. His sister Alice's diaries, published after her death, were widely read as an example of early feminist thinking. Religion was an essential part of James's early life. His father's view of its importance never left him. His own life, however, turned in the direction of science. After receiving a medical degree at Harvard, James soon turned his attention to psychology, where his work had enormous influence. His two-volume textbook, *Principles of Psychology*, became a standard in the field. It was so omnipresent in the early years of psychology that the volumes were given nicknames by the generations of students who grew up with them. The single-volume short course was called "the Jimmy," while the full two-volume set was called "the James." But as James reached his middle years he found his interest turning from pure psychology to wider issues of both philosophy and religion. When the invitation came to give the Gifford Lectures James was ready.

The Scottish judge Adam Gifford had endowed the lecture series bearing his name a year before his death in 1887. The "Giffords," as they came to be called, were to be held at Scotland's great universities. Their purpose was to stimulate "lively and perpetual debate on science and 'all questions about man's conception of God and the infinite.'"[2] Gifford hoped to bring the world's great thinkers to Scotland to address the growing gap between science and religion. By 1902 this hope had been realized. After only a decade the Gifford lecture series already drew respect from scholars on both sides of the Atlantic. For William James the invitation would make it possible to reconcile his scientific thinking with his need to understand questions he inherited from his father.

James stayed away from grand overarching theories that tried to explain everything under a single rubric. He was suspicious of academic theology. The systematic "block universes" they created were, to him, sterile inventions of the mind and did not touch the real importance of spiritual feeling. With this perspective he began his yearlong journey to Edinburgh. In the wake of his father's death he wrote of his desire "to understand a little more of the value and meaning of religion in [his] father's sense, in the mental life and destiny of man." As he traveled across Europe that year he turned his questions in the only direction that made sense to him. Instead of addressing theology and academic definitions of God, James claimed the high ground of the empiricist. He turned directly to the role of experience in religion. For James, this must be the focus of any psychological account of human life's sacred dimension. Word spread, and people came: students, professors, ministers, agnostics. The audience grew larger until, finally, there was no space left in the lecture hall.

WILLIAM JAMES AND
THE PRIMACY OF EXPERIENCE

One year after his Gifford Lectures, James gathered them in a book, *The Varieties of Religious Experience*. It quickly became a classic in the study of religion. It remains a staple of the curriculum of religious studies departments across the United States. As one critic describes the work, "It is our best book about religious experience, our best defense against skeptics, and our surest incitement to a genuine public dialogue about the significance of personal religious experience for our common life."[3] In our own search for a different and more enlivened perspective on science and religion, James's focus on experience offers us a kind of clearing in the tangled forest of debate. It gives us a different path through the woods.

What makes the *Varieties* such a milestone in the discussion of religion is James's cheerful willingness to bypass traditional debates and see spiritual life from a fresh and original perspective. The experiences he uses are those of other people. He takes written records, published accounts, diaries, and memoirs as his raw material. James acts as a keen, fairly detached observer of his subjects' experiences. Simultaneously he allows himself to participate in the enthusiasm those experiences engender. Holding those opposing positions is the tension that animates *The Varieties of Religious Experience*. It is a tension we must also balance as we attempt to find a more encompassing view of science and spiritual life.

Early in the book James offers his now-famous definition of religion:

> Religion . . . shall mean for us the feelings, acts, and experiences of individual men in their solitude; so far as they apprehend themselves to stand in relation to whatever they may consider the divine.[4]

This perspective on religion stands in stark contrast to the idea of religion that traditional, popularized debates over science and religion focus on. Here the emphasis on solitude is crucial. It reflects James's wish to understand an elemental encounter with perceived spiritual realities. He is not interested in theological theories. As he writes, "The problem I have set myself is a hard one: . . . to defend . . . 'experience' against 'philosophy' as being the real backbone of religious life."[5] That turn from theology to experience irrevocably alters the character of the inquiry and the nature of questions James asks us to address. To begin with, the elemental encounter with life's sacred character must be distinguished from the derivative life of rote religious practice. James makes this distinction quite clearly as he separates the authentic religious experience from the "ordinary religious believer, who follows the conventional observances of his country."[6] It is experience, not institutional practice, that is primary for William James. Experience, he claims, stands alone as the root of every established religion. "Personal religion will prove itself more fundamental than . . . theology," James writes. "Churches, when once established, live at second hand upon tradition but the founders of every church owed their power originally to the fact of their direct personal communion with the divine."[7]

The divine is a term that many scientists blanch at, but James gives considerable latitude in his definition of the term. Most important, he does not force it into the concept of God in the Abrahamic tradition. James explains, "A chance of controversy comes up over the word divine if we take the definition in too narrow a sense. There are systems of thought which the world calls religious and yet do not positively assume a God. Buddhism is in this case." Later James adds, "Accordingly, when in our definition of religion we speak of an individual's relation to what he considers the divine, we must interpret the divine very broadly."[8] What matters for James is not a person's experience of God in a scriptural sense. Instead, religious experience describes the individual sense of an encounter with the source of the sacred. It is an encounter with the sacred character of the world *as it is experienced*. That experience may be interpreted in the context of a particular religious tradition, but it need not be.

A crucial distinction can be made when one turns from religious doctrine to religious experience: the potential to relinquish the supernatural. When one focuses on the ways in which people, ordinary people and saints alike, become aware that there is a sacred character of Being, then the discussion changes. The science and religion debate can finally turn away from comparing the supernatural powers of a Creator with the timeless laws of nature.

James is interested in the experiences people have in solitude. What these experiences bring to people's lives is what matters, not what they reveal about an abstract idea of God's powers. In his emphasis on religious experience James opens the door to us for the consideration that spiritual experience and spiritual life can occur within the field of the natural world with its wholly natural accounts of the world's order. What James will ask, however, is that we see these experiences as irreducible. That is, we must understand that they live in the immediacy of the moment and cannot be reduced to simple accounts or descriptions.

It is one thing to explain the world as an object; it is another to experience it as a subject.

James positions himself as a psychologist, a practitioner of science, but he does not doubt that the experiences he relates have content. There *is* something to them, he tells us. There is some truth that is revealed. James dismisses what he calls "medical materialism" as an explanation for religious experiences. They cannot be swept away as conditions of an overwrought nervous system or reduced simply to neurology.[9] Follow such attitudes to their end, he says, and all beliefs, all opinions, would be reduced to biology. If such a dismissive view of religious experience were true, he quips, then "we should doubtless see 'the liver' determining the dicta of the sturdy atheist just as decisively."[10] In other words, if a profound religious experience is dismissed as the work of a spoiled turkey sandwich, then the thoughts that lead to its rejection might be just as easily discounted.

What exactly does James mean by "religious experience"? In the *Varieties* he quotes from sources describing a diversity of encounters with the world's sacred quality. Some of his accounts come from well-known figures in the history of religion, including Saint Teresa of Avila and George Fox, the founder of Quakerism. He also draws from people who stand firmly in the clearing of their own lives but are not recognized by history. These descriptions of deep and intense experiences from ordinary people give the *Varieties* its particular emphasis. We hear anonymous voices speaking of "the temporary obliteration of the conventionalities which usually surround and cover my life" and "What I felt on these occasions was a . . . loss of my own identity, accompanied by an illumination which revealed to me a deeper significance than I had been wont to attach to life."[11] Though James is drawing his accounts from sources who, for the most part, live in "Christendom," many of the descriptions are not about God or the divine in a theistic sense. Instead they cover a broad spectrum of responses to the lived sense of an unseen order. They are responses to that which constitutes something "more," something beyond the pale of the ordinary.

Many of the experiences related in the hundred-year-old *Varieties* have a distinctly quaint, Victorian quality. To understand what William James means by "religious experience" a modern description may be more compelling. For this we can turn to Daniel Quinn. Quinn describes an archetypal religious experience that occurred during his brief training as a Trappist monk. As he stepped outside on a fall morning to begin work in the monastery gardens, Quinn was overtaken by a vision of the world in its full clarity:

> I turned and faced the sunshine and the breath went out of me as if someone had punched me in the stomach. That was the effect of receiving this sight, of seeing the world as it is. I was astonished, bowled over, dumbfounded.
>
> I could say that the world was transformed before my very eyes but that wasn't it — and I knew that wasn't it. The world hadn't been transformed at all; I was simply being allowed to see it the way it is all the time. I, not the world, was transformed.
>
> Everything was burning. Every blade of grass, every single leaf of every tree was radiant, was blazing — incandescent with a raging power that was unmistakably divine.[12]

Quinn's experience was transformative. He describes it as the most important hour of his life. His encounter that morning changed his understanding of himself and the life he was surrounded by. His experience in the garden became the basis for years of effort as he sought to understand what he had "seen." The effort culminated in *Ishmael*, the novel for which he won one million dollars and the 1991 Ted Turner Tomorrow Fellowship. It continues to sell copies and draw followers.

William James would have recognized what Daniel Quinn went through that day in the monastery garden. He also would have recognized the effort that followed from it. James spends some time in the *Varieties* on the transformative power of these encounters with the lived depths of spiritual experience. People feel changed by their experience the way Quinn felt changed. This follows because the domain of religious experience is an interior one. In its exploration our sense of ourselves and of the world can be reshaped.

The response to experience represents an important departure point for us in our attempt to forge a different understanding of science and spiritual endeavor. Religious experience often drives an aspiration to make more complete contact with its perceived source. This aspiration then leads people into what James calls the "strenuous life." As a psychologist James saw the potential power that can emanate from such experiences. He writes, "That prayer or inner communion . . . with the spirit . . . is a process wherein work is really done." Here James points to the capacity of religious experience to push people into the realms of effort. Such effort can be directed at the inner work of deeper reverence for the world or gratitude for its gifts. It can also be directed outward in compassionate action. It can also simply manifest as a desire to know more. Religious experience can drive an aspiration to act, an effort to search more deeply. Here we can find parallels in the aspirations that drive efforts in science both individually and for the culture as a whole. Science is also a means for knowing the world more clearly. The elemental effort it requires can also derive from a lived sense of the world's great beauty and mystery.

Much of the *Varieties* is devoted to articulating the different forms religious experience can take: a sense of health and well-being; the urge to penitence or saintliness; the impulse to conversion. As he lays out these varieties of experience James makes it clear that he is after universals. He wants to know what is common to all people, independent of creed or calling. When he finds commonality in his accounts, James claims that its existence points to deeper connections between the different, established forms of religion. Since, for him, the core of all religions lies in experience, James concludes that the essence of all religions is the same. In this way James is a religious pluralist. What lies beneath the different forms of institutional religion is the universal experience of the world's spiritual quality, its sacred character. "I am expressly trying," he writes, "to reduce religion to its lowest admissible terms, to that minimum, . . . which all religions contain as their nucleus, and on which it may be hoped that all religious persons may agree."[13] This is a useful perspective for us as we struggle to free ourselves from the tyranny of results-based comparisons of science and religion. We *need* to find what can be drawn together from the breadth of experiences people call religious.

One commonality that is of interest to us is what James calls the "reality of the unseen." As he describes it, "It is as if there were, in the human consciousness, a *sense of reality, a feeling of objective presence, a perception* of what we may call *'something there,'* more deep and more general than any of the special and particular 'senses.' "[14] James refers to this as "an undifferentiated sense of reality." It is the sense of a "more," as he calls it. James takes this sense to be an essential aspect of religious experience predating growth into any particular established religion. "[The various theologies] all agree that the 'more' really exists; though some of them hold it to exist in the shape of a personal god or gods, while others are satisfied to conceive it as a stream of ideal tendency embedded in the external structure of the world."[15] It is noteworthy that in Daniel Quinn's experience this "something more" appears as the world itself revealed without the blanket of Quinn's normal perception. For Quinn the "more" is always there but only becomes apparent during his moment of religious experience.

Another commonality of religious experiences is their so-called noetic, or knowing, quality. Some truth about the world has been acquired through the experience. Equally important, these experiences have a lasting effect. They change people. There is a deeply felt aspiration to understand and live in accordance with the perceived reality they felt in contact with. For James, the noesis, the knowing that comes in religious feeling, and the changes in conduct it engenders are far more primary than the ideas and concepts of different religions:

> When we survey the whole field of religion, we find a greater
> variety in the thoughts that have prevailed there; but the feelings
> on the one hand and the conduct on the other are almost always

the same, for the Stoic, Christian, and Buddhist saints are practically indistinguishable in their lives. The theories which Religion generates, being thus variable, are secondary; and if you wish to grasp her essence, you must look to the feelings and the conduct as being the more constant elements.[16]

Experience is what matters because that is where effect and affect are manifested. James insists that religious experience, rather than religious belief, is the true foundation for a pluralist view of religion as a human phenomenon. In doing so he provides us with a vital stepping-stone as we seek an alternative view of science and religion. The capacity to hold religious experience as a whole human phenomenon rather than have it fragment into various institutional forms is a critical step in changing the nature of the discussion. If religious experience is at the heart of religion and if there are common themes in religious experience, then we can turn away from the claims of one institutional religion versus another. We can, instead, ask what gives these experiences their power and what connection they have with the activities and efficacies of science. When James points to common themes such as the reality of the unseen and a noetic capacity to understand something of the world we can then ask if, and how, these connect to similar themes and experiences in science. Doesn't science reveal unseen structures of the world, from electromagnetic fields to information transfer in DNA? Isn't science about the development of understanding and knowledge?

But we must be careful. The point here is explicitly *not* to claim that something identical occurs in both science and spiritual endeavor. Rather it is to say that they are parallel in aspiration, with similar effects and affects.

The concept of religious experience forms a natural means for redirecting our attention in the debate between science and religion. William James brought the idea of religious experience into the mainstream by setting it in the broadest possible context and framing it within psychology. But the idea did not begin with James, and it did not end with him. In both its origins and continuation we will find more raw material for building a bridge with science and gaining higher ground.

EXPERIENCE IN DIFFERENT KEYS

We have already seen how the Enlightenment marked a transition in the cultural role of religion. During the late eighteenth and early nineteenth centuries the growing desire to throw off the intellectual constraints of the past led many to bridle under the conventions of established churches. During this same time science was rising into its place of cultural preeminence. It was in this setting that religious experience as a separate way of thinking about religion was first given voice. It was no coincidence that the rise of science and the rise of religious experience as an idea occurred at the same time. As Wayne Proudfoot, a scholar of religion, puts it, "The idea of religious experience [was] shaped by the conflict between religion and the growth of scientific knowledge."[17]

It was Friedrich Schleiermacher (1768–1834), a German philosopher and Reform preacher, who first championed religious experience as an idea in and of itself. In Schleiermacher's two most important books, *On Religion: Speeches to Its Cultured Despisers* and *The Christian Faith*, he sought to give religion a new argument against the growing number of its Enlightenment critics. Schleiermacher moved in a circle of Berlin writers and artists who were impatient with traditional religion and saw it only as a fetter. By focusing on its experiential nature and grounding it in a direct encounter with the world, Schleiermacher hoped to show his friends that religion could also be freed from the weight of institutions and prejudice. As Proudfoot describes it:

> Schleiermacher offered a careful description of the religious sense from the perspective of a member of the communities with which he was most familiar. He sought to convince his friends among the artists, poets and critics of Berlin . . . that their sensibilities were

more in tune with the genuine spirit of religious life than much of
what went on in the churches and synagogues. He set out to awaken
this interest and direct their attention to it.[18]

The idea of religious experience would not have occurred as such in
theology before Schleiermacher. Religious thinkers did not conceive
of religion in such terms because there was no need to. It was only in
light of the emergence of science as a cultural power, and the hope
of liberation tied to that emergence, that established religion became
seen as the oppressor. Only at this point did religion need cultural and
cultured advocates. Against that background Schleiermacher's evocation of "experience" as a separate way of understanding human spiritual endeavor marked a turning point. By making religion a deeply felt
quality of lived experience, Schleiermacher connected it to the same
liberation his friends found in their work as artists and poets. Proudfoot
explains:

> The turn to religious experience was motivated in large measure
> by an interest in freeing religious doctrine from dependence on
> metaphysical beliefs and ecclesiastical institutions. . . . This was
> the explicit aim of Schleiermacher's *On Religion*, the most influen-
> tial statement and defense of the autonomy of Religious experi-
> ence. . . . Religion had its own integrity and religious belief and
> practice were properly viewed as expressions of the religious
> dimension or moment.[19]

Schleiermacher's aims are similar to ours in the sense that we are
interested in freeing the debate between science and religion from its
own bonds in the present day. In a very real sense those of us today who
wish to preserve the power and efficacy of science and still retain the
reality of our inmost spiritual experience can turn in the same direction. What if science and religion turned not on what each says but on
what each evokes in the domain of experience? To see how this might
work we will have to press on.

One hundred years after Schleiermacher and long after science had established its claims on truth, another German philosopher and theologian would leave his stamp on the notion of experience. In 1917 Rudolf Otto published *Das Haulage*, translated as *The Idea of the Holy* (it is noteworthy that the title could also be translated as *The Idea of the Sacred*). This short book, like James's *The Varieties of Religious Experience*, remains a staple of religious studies courses. In *The Idea of the Holy* Otto uses an exacting descriptive scalpel to analyze the quality of the most intense spiritual experience — the direct confrontation with the sacred. In prose typical of German philosophers, Otto slices away distinctions in micro-thin layers as he makes a "serious attempt to analyze all the more exactly the feeling that remains where the concept fails."[20]

In the deepest experience of the sacred, Otto claims, we encounter a quality he calls the "numinous." For Otto the numinous is more of a mental state than a feeling, occurring as the most elementary level of spiritual experience. Pinning down the numinous in words is problematic: "like every absolutely primary and elementary datum, while it admits of being discussed it can not be defined." The only way to help someone understand the numinous, according to Otto, is to guide them in its consideration until "it begins to stir, to start into life."[21] Otto's simultaneous evocation and articulation of the numinous is rich and cannot be done justice here. He draws attention to experiences that are akin to that of the numinous and more easily recognized, for example, the experience of being swept away by an especially beautiful piece of music. All of us have been carried into states of heightened sensitivities through song, whether it be Beethoven's *Ode to Joy*, John Coltrane's "A Love Supreme," or Led Zeppelin's "Whole Lotta Love." There is an intensity in the experience of music that is more than simply feeling. But for all their power, Otto says, these experiences are still poorer cousins to the quality of absorption and transport that come through encounters with what we apprehend as sacred.[22]

Otto's prose, in translation at least, can be turgid, but the depths of his commitment to articulate encounters with the numinous is inspiring. As the twentieth century matured many scholars of religion drew on Otto, including the influential University of Chicago doyen Mircea Eliade. For our purposes Otto's attempts to provide an incisive description of the deepest religious experiences sets the stage for our next step, which is to identify the location of our encounters with life's sacred character and use it to see science and spiritual endeavor as active, complementary parallels. Otto also presents us with a continuum of emphasis, along with James and Schleiermacher, on experience over doctrine as the source of religious understanding. This is the bridge over which we can cross between the domains of science and spiritual endeavor.

EXPERIENCE UNDER FIRE

Before we go further, the good scientist in me demands a message from our sponsor, the human faculty of critical thinking. Given the importance of shifting the science and religion debate from doctrine and results to the new domains of experience, we would do well here to pause for a word of caution. While a focus on religious experience will take us a long way toward building a different and renewed perspective, it is not a concept without its detractors. It is too easy in the domains of science and religion to fall prey to claims of grand correlations and correspondences. If we want to keep to the spirit of both science and true spiritual endeavor we must not abandon our capacity to doubt.

In his book *Religious Experience*, Wayne Proudfoot tracks the emergence and development of this concept from Schleiermacher onward. Proudfoot has no problem with seeing religious experience as a useful tool for thinking about religion. The fundamental insight it yields, according to Proudfoot, is the insistence that *the subject*, the person having the experience, is what matters. In his words, "To identify an experience from a perspective other than the subject would be to misidentify it."[23]

The problem, according to Proudfoot, is the claim that these experiences are so special that they cannot be judged outside of "spiritual realities." He calls such an approach "protective": "Schleiermacher's claim that religious experience is independent of concepts and beliefs functions as a protective strategy. It precludes any conflict between religious belief and the results of scientific inquiry."[24] In Proudfoot's view, Schleiermacher, James, and Otto are all trying to circle the conceptual wagons around religious experience and keep it safe from skeptics.

The skeptic, of course, would argue that no matter what a person "feels" there is nothing real, nothing truly "more" that exists as the source of religious experience. The protective strategy of James and others is, according to Proudfoot, designed to shield religious experience from *explanations*. Recall the spoiled-turkey-sandwich dismissal of a profound religious experience. Proudfoot tells us to be suspicious of any attempts to exclude it and all nonreligious explanations from religious experience. In short, the protective strategy allows the religious explanation of religious experience to go unchallenged.

Explanation *is* the heart of the issue. James, Schleiermacher, and Otto all claim that religious experience is, somehow, original and fundamental and does not depend on a set of beliefs or ideas. James argues that all established religions derive from the more elemental domain of religious experience. Religious experience can, in this view, be separated from any prior thinking or belief about religion. Proudfoot does not buy any of this. No such separation is possible, he claims. Experience is always conditioned by what we bring to it: "The term religious, when used to modify experience, is not independent of beliefs and explanations but assumes a particular kind of explanation."[25] What Proudfoot means is that a religious person wants religious explanations for his or her religious experience but a skeptic would be quite happy with the spoiled-turkey-sandwich explanation. In both cases explanation and background are required. You cannot have an experience that you call religious if you do not have some idea of what religion or spirituality means.

Do Proudfoot's arguments close the door on our use of religious experience as the basis for getting a different, and better, view of science and religion? They do not; but his claims against the priority of religious experience must be considered for two reasons. First, the point of science and authentic spiritual practice is to not fool yourself. Thus we need to keep our critical eyes open. Second, the argument against a protective strategy for religious experience forms perhaps the only part of the results-based debate between science and religion that is inherently vital: the program of neuroscience.

As neuroscience advances there are many who claim that all experience, religious or otherwise, is ultimately reducible to brain states. The ecstatic religious states that precede epileptic seizures have long been documented as an example of a so-called sacred experience that has a definite neurological cause. More recently a number of researchers have claimed to have found regions of the brain that activate religious feeling. What if they are right? What if those who see *all* aspects of consciousness, all aspects of being a subject, as solely neurochemical are right? What happens to the special role of religious experience then?

Although I watch the advances in neuroscience with great interest, I suspect that ultimately they can produce only a fraction of the truth. Mostly I believe this because it is not clear to me that science is amenable to telling this particular story in its entirety. You could hook me up to an MRI forever and locate all the parts of my brain that light up when I bite into a crisp apple, but that would still not exhaust the content of my experience. Do I have emotions, for example, get angry, because a certain part of my brain lights up, or does a certain part of my brain light up because I get angry? There is something irreducible about being a *subject* that science does not yet have words for. There is something about *existing*, about the strange verb *to be*, that cannot be accounted for with words, graphs, or explanations. It must be lived. There is a gap between our explanations for phenomena (for all their power) and the fundamental presence of those phenomena as the field of our experience. I argue that Being remains, at its root, remarkably and wonderfully open-ended. So, for my part, from my experience, I agree with the James gang that experience is primary. It is the immediacy of experience, in particular the experiences we call spiritual, that gives it such poetry and power in human life. While we will need to keep Proudfoot's dissent in mind, it seems to me that mere intellectual argument cannot wither the strength of a tree that has borne such potent fruit for so many people over so long a time.

THE SACRED AS EXPERIENCE

Language limits us, and language illuminates the path forward. In imagining a different relationship between science and religion our greatest challenge lies in finding the right words. Science strives for precision to corral and delimit the stuff of its investigations. A lot of work goes into specifying the meaning of mass, energy, planet, virus, cell, ribosome, and so on. The content of spiritual endeavor is not so easily captured. Otto told us that the elementary expression of religious experience can be discussed but not defined. It is a point well taken. Given this disparity between the domains in the usefulness of definitions, we can expect problems to arise in attempts to find their proper relationship. We must choose our words carefully. For a scientist talking about the domains of human spirituality there are a host of terms that present difficulties: God, the divine, and the supernatural, among others. Anything that undercuts the capacity for science to move unfettered in its articulation of the world's structure, any metaphysical tap-dancing that articulates invisible angels dancing on pin-heads, bleeds the integrity from the discussion. What words can retain the living, experienced quality of spiritual life without drawing us back into a debate that has failed to give us a bridge between the two domains?

I have deliberately relied on the term *sacred*. It is rich with resonance without being connected to any particular religion. It has a meaning that is intuitively apprehended without specific definitions and is relatively free of the connotations and baggage of terms such as *God* and *divinity*. Most of all it seems to strike a fundamental tone that gives religious experiences their religious character. As the *Encyclopedia of Religion* defines it, *the sacred* is "what gives birth to religions, in that man encounters it; or it functions as the essence, the focus, the all important element in religion." In *sacred*, then, I believe we have a word and an idea that cuts both deep and broad. Does what we call sacred have any connections with what happens in science? I believe the answer is yes, as long as we do not turn the experience of life's sacred character into *the sacred* as a substitute for *divinity*, and as long as the emphasis remains on what we encounter in experience. Can we unpack "the sacred" and use it to frame a new view of science and religion?

Just as our exploration of religious experience took us to William James, traveling into the territories of the sacred will take us straight to the door of Mircea Eliade. Eliade was a scholar of prodigious output and interests, a character in his own right. A Romanian by birth, he journeyed to India as a young student in 1928 and carried out groundbreaking research on the structure and meaning of yoga. During World War II he served with the Romanian Legation in Britain and Portugal. After spending time in Paris, he took a position at the University of Chicago and for years was widely regarded as a leading scholar in the study of religion. It was partly through Eliade that the Chicago School of Religion gained dominance. Throughout his life he was a controversial figure and a lightning rod for both praise and criticism (especially for his right-wing views espoused in the 1930s). Eliade was tireless, publishing more than one hundred works of scholarship and fiction. His book *The Sacred and the Profane* remains a widely read, though contentious, work in religious studies on a par with James's *The Varieties of Religious Experience* and Otto's *The Idea of the Holy*.

Eliade drew on the "sacred" because it is pregnant with possibilities. The origin of the term tells us something about its use for Eliade and the way it can serve as a fulcrum to leverage our own discussions of science and religion. It is related to the Latin *sacrum* — "what belonged to the gods or what was in their power."[26] Its early usage was reserved for Roman temples and their rites. In that context the words *sacrum* and *profanum* were paired. The profanum was the space in front of the temple, outside. The sacrum was the inside. "A spot referred to as *sacer* was either walled off or otherwise set apart — that is to say, sanctum — within the surrounding space available for profane use."[27]

Here we find a compelling resonance. The sacred relates back to a specific location, a space and a time, set apart from the day-to-day happenings of life. The commerce, contest, and competition of the ordinary world occurred outside in the profanum. Inside, within the sacer, humans entered a realm of a different order. For the Romans it was the realm of their divinities. There they were expected to be responsive to divine needs in both behavior and attention. Given these origins, the terms *sacred* and *profane* have remained paired in the study of religion.

Separating the sacred from the profane was crucial for Eliade. It guided his thinking about religion and what he called religious man. Eliade begins *The Sacred and the Profane* with a nod to Otto and *The Idea of the Holy*. "Gifted with great psychological subtlety," Eliade begins, "[Otto] succeeded in determining the content and specific characteristics of religious experience." Eliade, like Otto, points to the numinous, that elusive but illuminating ground of religious experience: "The numinous presents itself as something wholly other, something basically and totally different." This is how Eliade launches his account of religious man's confrontation with the sacred. We exist in the midst of two worlds: "The sacred always manifests itself as a reality of a wholly different order from 'natural' realities."[28]

This sense of the "wholly other" is what appears directly in our experience. What matters is that we encounter the sacred. It appears, or erupts, into our lives. Thinking about it, theorizing about it, misses its essential, living power. Eliade is explicit. There is a fundamental gap that language cannot cross. The experience cannot be distilled to definitions or analytic concepts. Simply put, words fail. Eliade writes, "Language is reduced to *suggesting* by terms taken from that experience."[29] The experience of the world's sacred character cannot be wrapped up and contained; it can only be pointed to through metaphor or analogy. Eliade himself does not define the sacred. He seems to infer it is not possible.[30] Instead he tries to show how the experience of the sacred forms the wellspring from which spiritual endeavor and religious life emanate.

Eliade offers a new term, *hierophany*, to designate where and when the sacred erupts into the world. A hierophany is the manifestation of the sacred, the act of its appearance in the world. It occurs when "something sacred shows itself." According to Eliade, this process forms the heart of all religious life: "It could be said that the history of religions — from the most primitive to the most highly developed — is constituted by a great number of hierophanies, by manifestations of sacred realities."[31] The hierophany is the capacity of the sacred to appear in the midst of the profane.

Most of us have experienced this. Perhaps during a walk in the woods, a period of solitude, or a time of intense concentration the world feels utterly different, more intense and more present. This can be a hierophany. For Eliade, however, the hierophany as an idea took him to a deeper aspect of religious man's existence. It pointed to something that could become more established and more intentional. The hierophany, according to Eliade, was once an elemental part of culture. This experience of sacredness, of the world standing out on its own, is so powerful that early on human beings began to seek it out and nurture its appearance. It is of no small importance for us in our search for a broader perspective on science and spiritual endeavor that Eliade conceives of the hierophanies as stretching back to our earliest mythologies and religions.

In archaic cultures a tree in a glade or a stone in a field could be taken as a hierophany. The site might have been selected through experience or through a sign such as a dream. What matters for Eliade is not the nature of the specific object: "The sacred tree, the sacred stone are not adored as a tree or stone; they are worshipped precisely because they are hierophanies, because they show something that is no longer stone or tree but the sacred."[32]

An example will illuminate the idea. The Khanty tribes of Siberia have lived in the lowland river basins of Russia's western frontier for thousands of years.[33] The Khanty are largely a nomadic people whose life of hunting and fishing lasted well into the twentieth century. Despite the best efforts of the Orthodox Church and the Soviet regime, traditional Khanty belief and ritual have managed to survive. Ethnologists attempting to understand traditional Khanty ways before they are lost to modernization have found these nomads treating landforms as sacred in exactly the way Eliade described. High places are relatively rare in the Khanty world, so hills, hillocks, and ridges become hierophanies. The Khanty have a special name for these locations: the Kot Myx. A Kot Myx, or God House, is essentially a glade or an opening in the forested hill that may include a tree deemed especially sacred. Newly married Khanty couples come to a Kot Myx to pray, make offerings, and ask for blessings at the beginning of their life together. Like the interior of the Roman temple, each Kot Myx is a sanctum and a sanctuary. Khanty taboo prohibits the removal of anything from the site of a Kot Myx. These "ordinary" glades are not ordinary to the Khanty, who see them transformed into a sacred space. They are gateways allowing Khanty tribespeople to experience the wholly other, wholly sacred character of the world. Upon entering these spaces the world becomes numinous. Space and time are experienced differently than in the profane world of the everyday.

The description of the Kot Myx should not be unfamiliar to us in the modern world for we have our own versions of hierophanies. As human society became more organized and complex, the location and nature of these sacred sites changed; instead of glades, sacred sites became temples, basilicas, and cathedrals. Like the entrance to a Kot Myx glade, the church door serves as a threshold separating the noise and chaos of the profane from the quiet solitude of the sacred.

Ironically, as cities became vast metropolitan complexes, many people sought encounters with the sacred, their hierophanies, in nature, just as the Khanty do. Although we are still capable of encounters with the sacred, according to Eliade, something has changed in our "ascent" to modernism. Something has been lost.

In articulating pathways of the sacred and profane, Eliade looks forward as well as back. After millennia of efforts to draw closer to the sacred, Eliade claims, the world has become *desacralized*. It has become fully and completely profane. According to Eliade, religious man of archaic societies tried to live in close proximity to the sacred. It could appear in all vital functions of life: food, sex, work. Hierophanies took diverse forms, and their presence was stitched into the fabric of everyday life. Now, for the most part, that fabric has been torn and our capacity for communion with the sacred has diminished. Eliade writes, "Simply calling to mind what the city or the house, nature, tools, or work have become for the modern or nonreligious man will show with the utmost vividness all that distinguishes [him] from a man belonging to an archaic society or even a peasant from Christian Europe."[34] In Eliade's view, a kind of cultural discontinuity occurred over the past few hundred years. It tore us from a world whose sacred character was always close at hand. The modern world has become unbalanced. "It should be said at once that the completely profane world, the wholly desacralized cosmos, is a recent discovery in the history of the human spirit."[35]

Many people would point to science as the root cause for the desacralization of the modern world. Many people, including perhaps Eliade, would see the domination of science as the prime mover in the great gap that opened between humanity and its original aspiration. This need not be the case. When speaking of the relation between science and religion, the biologist Ursula Goodenough once observed, "To assign attributes to Mystery is to disenchant it, to take away its luminescence."[36] Goodenough wanted to express the possibility that there is room in science for Mystery. She might also have been speaking of the possibility that there is room in science for the sacred.

SCIENCE AND THE SACRED

When I heard the Learn'd Astronomer;
When the proofs, the figures, were ranged in columns
    before me;
When I was shown the charts and the diagrams, to add,
    divide, and measure them;
When I, sitting, heard the astronomer, where he lectured
    with much applause in the lecture-room,
How soon, unaccountable, I became tired and sick;
Till rising and gliding out, I wander'd off by myself,
In the mystical moist night-air, and from time to time,
Look'd up in perfect silence at the stars.

Walt Whitman

I understand Whitman's sentiments, but this poem always makes me sad. I have heard these same feelings many times from friends and students. For them the veneer cannot be cracked and the world science reveals seems as dead as specimens stored in formaldehyde. I wish someone else could have shown Whitman what lies behind the proofs, what hovers like a mist unseen around the charts and diagrams. There is a different kind of poetry emanating from the moist night air, and it can be read clearly in the language of science.

Whitman's poem describes a spiritual response to a desacralized cosmos. As Whitman describes it, science disenchants the world, killing its spiritual quality through relentless logic and the tyranny of "brass instruments" that measure, meter, and master the world. Whitman's potent images of sickness and exhaustion in his encounter with this dead cosmos are balanced by the renewal he feels in the presence of the sacred, "the mystical moist night-air" and the "perfect silence" of the stars. Whitman might not see science as providing any links to spiritual endeavor, but the language we have learned in this chapter opens new doors and, perhaps, provides us with another way of speaking about science. Eliade's desacralized cosmos is not the only place a culture dominated by science need inhabit.

By taking what William James calls religious experience as a focus we can change the science and religion debate. This provides the first terms in a vocabulary that can speak about the two enterprises with a poetry that illuminates both. The focus on experience that James, Otto, and others have articulated is a significant departure from the emphasis on creed and scripture that many of us associate with religion. If religion is instead about "religious experience," then we can turn in a different direction as we seek parallels with and bridges to science. Ideas of God are secondary to the experience of the more that occurs again and again on the level of the personal in every culture and every time. This is something that science and scientists cannot and should not deny. Experience frees us from the imperatives of any particular system of belief. Its claim is one of universals. The focus on experience allows for us to see that even in the solitude of our interior life there arises a commonality that transcends cultural distinctions. In that transcendence we can find a new way to describe science and spiritual endeavor.

We do not have to compare the latest results of cosmology and evolution with someone's scripture. We do not have to compare the Heisenberg uncertainty principle with Buddhism. Instead, we can speak of heartfelt yearning and the longing that grows from experience of the lived mystery lying at the heart of Being. For some this is directed to an explicit more of the kind James describes. For others, like the scientist Ursula Goodenough, leaving it as mystery pure and simple is enough. In either case we acknowledge that being human involves an encounter with a quality we call sacred. These experiences have a noetic quality; something about the world is known that was not seen before. Religious or spiritual experience, like the endeavor of science, responds to the world's beauty with an open-ended question: What is this? Such knowing is the "still small voice" heard and known with an intensity that leaves an indelible mark.

THE MYTH OF FIONN: VOICE OF THE WORLD

The firelight spreads a dancing ocher glow around the encampment. The traveling band of hunter-warriors is in a fine mood after a long day in the Irish wood. The men joke and laugh but always keep an eye to their leader, the great hero Fionn McCool. He is their stalwart, their paragon. Each of the men scattered around the fire pit is an army in his own right, and few in the Irish kingdom would be foolish enough to go against any one of them, but Fionn is their first. The men look to him for his strength and great wisdom.

Fionn (pronounced Finn) was born to lead, but his path through life had not been an easy one. His father, Conn, had been the leader of the Fianna, the warriors who would aid Ireland's kings in times of war. Fionn's mother was the beautiful Muirne, daughter of a Druid high priest. The love of Conn and Muirne had inflamed the girl's father, and, in the end, Conn was killed in the battles that followed his attempts to wed Muirne. Already pregnant when Conn died, the Druid girl fled. When the boy was born she left him to be raised in the secret forests, unseen and unknown to any of the Fianna tribe. There two women, the Druidess Bodhmal and the warrior priestess Liath Luchre, taught the young boy the arts of war and hunting.

The boy grew quick and strong with a mind for magic and wisdom. After serving under a series of kings the young Fionn met the wizard Finnegal. The diminutive Druid schooled the young man in Secrets. In time Finnegal came to trust his young charge enough to bring him along in the great quest for the fish of knowledge. The prize was nothing less than complete understanding of the world's sacred ways, for he who ate of the salmon's flesh would gain all that could be known of the universe. When, after many years, the Druid finally caught the great fish he asked Fionn to cook it. Dutifully obeying his master's order, Fionn promptly burned his thumb on its skin. Unconsciously taking his thumb to his lips to salve the burn, Fionn gained some part of the wisdom imparted in the great fish. The knowledge and understanding Fionn gained that night became a weapon more potent than his strength and courage in the battles awaiting him.

In time Fionn regained the title his father had rightly held and was hailed as the greatest leader the Fianna had ever known. He led with a bravery and compassion that made him beloved to those he served and feared by those he opposed. As the years passed Fionn became Ireland's greatest hero, a rival for all the world's Jasons.

But on this night, with the snap of fire-heated coals urging spark and smoke upward, Fionn's men think not of battle and bravery but of beauty.

"Father," asks Fionn's son, Oisin, "tell us what is the most beautiful music of the world." Fionn smiles and will not be drawn out so easily. "Perhaps it would be best if you tell me." "The bird call from the highest branch," replies the boy. "That is a good one," says Fionn, turning to the stout warrior Oscar. "And what think you, my overfed friend. What is the best music to your ears?" The large man thinks for a moment. "The top of all music is the ring of a spear against my shield!" he cries. "Aye," says Fionn, "that is great music indeed." Now all Fionn's men set to arguing. "The bellowing of a stag in the river," says one. "The laughter of a gleeful maiden," calls out another. "No, the whisper of a woman moved," responds a third. Fionn allows the clamor to continue for a bit and then stands, throwing his long coat of fur across his shoulders. For a moment no sound is heard but the fire and the movement of the treetops in the wind.

"These are all good sounds," says Fionn, looking to the trees and the stars above them. "But in truth there is one that you have all forgotten, the one true beauty in all the world. My dear companions, boon to my heart, always remember, it is the Sound of What Happens, that alone is the finest music in all the world."

. . .

Science is the careful and respectful consideration of this "Sound of what Happens." That is where it always begins, and that is where it always ends.

In this way science draws close to the domains of spiritual endeavor as they both grow from the immutable and irreducible fact of our experience.

Experience provides a language to frame a renewed perspective on science and spiritual endeavor. It is only from the experience of life's sacred character in the everyday world that we can identify a common ground. Eliade introduced the concept of hierophany to make the location and time when the sacred appears more explicit. Fionn's "Sound of What Happens" is the purest description of the way in which hierophany can wait for us behind every simple unattended moment. Now we can use the idea of hierophanies to see where and how science touches the potent energies of spiritual endeavor.

THE HIEROPHANIES OF SCIENCE

> Each new dawn is a miracle;
> Each new sky fills with beauty.
> Their testimony speaks to the whole world
> And reaches to the ends of the Earth.
>
> Psalm 19

Science, in its function and fruits, method and means, is a heirophany. It makes life's sacred character apparent to us. Science creates hierophanies through the act of careful observation and consideration. Science makes even the smallest thing — the gnat, the flea, the dust mote — sacred. This transformation comes about when the observer cares enough to notice the world's details. The experience of the star formation conference occurs again and again for scientists. It comes in the moment when we encounter a new image or recognize the key pattern in a new dataset. It comes whenever the pathways of science make the world stand out, illuminated and luminous. It comes when science shows us the music of what happens. It is a result of an encounter with the world when it is allowed to speak for itself. In hearing its voice we see the world as new and worthy of awe. It is true that many scientists might not identify this experience as such, but perhaps that is only because they too do not have the appropriate language to call it what it might become.

More important than the experiences of individual scientists is the relation of the culture as a whole. The view of the world that science provides has the power to transfix us. It can catapult our awareness out of the everyday, the profane, and into a space where nothing can be taken for granted. If, as a culture, we do not identify this moment as an encounter with the world's sacred character, as a hierophany, it is because we have been taught not to. Instead we call it "amazement," "wonder," or simply "awe." We should not be fooled. Rudolf Otto would not be fooled. In *The Idea of the Holy* he named "awe" as nothing less than the principal experience of the numinous.

The irony here is that in seeing science and its fruits as a hierophany we unpack Eliade's description and go beyond it. He told us that an experience of the sacred was an experience of the "wholly other" quality of the world. But from the perspective of science and spiritual endeavor the experience of life's sacred character is not an experience of some great "Other." It is not an experience of a reality not of this world. Instead it becomes the world viewed exactly as it is. It is the world experienced when we pay close, rapt attention. The hierophanies of science allow us to turn Eliade's ideas on their head. Taking a page from Zen, they show the sacred as not a thing in itself, a substitute for the abstract idea of divinity, but as the innermost and closest aspects of lived reality. The sacred in science is uniquely revealed because it emanates from the profane — the world directly before us if only we care to look.

This perspective is not new. Our cultural heritage gives us a perspective on science that limits our understanding of its deeper sources in the human imagination and spirit. If we remove these blinders science can become an encounter with James's "more" as what is really just before us in its own fundamental mystery. Science in its clearest experience can drive us out of the profane desacralized space and into a more open, attentive, and reverent one. Science can be a hierophany, a route to experience the world illuminated and sensed as sacred. Just recognizing that simple truth will take us a long way toward a vital and living relationship between the great endeavors of science and spiritual life.

CHAPTER 4

Not the God You Pray To

The Varieties of Scientists' Religious Experience

Science is piecemeal revelation.
Oliver Wendell Holmes,
The Works of
Oliver Wendell Holmes

Happy is he who gets to know the reasons for things.
Virgil

There can no great smoke arise, but there must be some fire.
John Lyly, *Euphues and Euphoebus*

Being a genius did not help. Wolfgang Pauli was tired and cold. He had wandered the winter streets of Zurich all night. It was January 1932, and the young physicist had reached bottom. To the world outside he was still a star, a vibrant, newly minted professor of theoretical physics at the prestigious Swiss Federal Institute of Technology. At just thirty-two years old he had proven himself a mathematical physicist of extraordinary depth even among the giants of his age — Einstein, Bohr, and Heisenberg. His demand for accuracy and clear reasoning earned him the nickname "the conscience of physics" from his colleagues.[1] His students saw him as a precision instrument boring through problems without effort. But intellect had its limits.

Pauli's problem lay in other arenas.[2] In the realm of his interior life, his "soul" (a word whose subtler meanings he would soon come to appreciate and use), he found only turmoil. No amount of academic success could assuage his pain. And his dreams . . . All his life he had been prone to vivid dreams, but now the internal theater had become dark and haunting. It left him exhausted in the morning. He could feel himself descending into crisis.

It had begun with his mother's suicide, an act of despair spurred by the revelation of her husband's affairs. Blinded by desperation, Pauli began a journey into darkness. In 1929 he hastily married a young dancer. The marriage was doomed before it began. After less than a year Pauli was divorced and his former wife was living with a chemist in Berlin. The rejection had been crushing. "If it had been a bullfighter," Pauli wrote a friend, "someone like this I could not have competed with — but such an average chemist!"[3] Drinking became a noticeable problem. On a 1931 trip to the University of Michigan Pauli tumbled, drunk, over a chair, breaking his shoulder.

Now, as the sun rose above the gray buildings of Zurich's downtown, Pauli's footsteps echoed his determination to break the grip of this crisis. At noon today he would face himself and the gathering storm of his dreams. He would have his first meeting with the great Swiss psychiatrist Carl Gustav Jung. Jung's radical ideas on the nature of the psyche and soul, of dreams and symbols and life's alchemy, had gained him a reputation in his own field as great as Einstein's was in physics. Contacting Jung was a radical move on Pauli's part and a measure of his desperation. Pauli was a scientist and rationalist to his core. To acknowledge the psyche could seem like an admission of weakness. But he had no choice. He had to face his interior life, the difficult pathways of the psyche and its openings to a deeper, more universal perspective.

As Pauli turned in the direction of Jung's office the wind was blowing hard from the east; the weather was set to change.

What began that day would stretch into a collaboration between the physicist and the psychiatrist lasting thirty years. Together they would explore the connections between the collective dreams of the human soul and the universal symbols of physics. What began as therapy would take Pauli on a lifelong exploration of his profound spiritual experiences.

· · ·

Science is a means for evoking the world's sacred character, a gateway to the experience of the world illuminated. This link between science and the experience of what is called sacred can take us a long way up the path. Before we journey any further we need to understand this relationship more fully. Wolfgang Pauli's journey from the realms of science to an exploration of his own, unique experience of spiritual life serves as a first example of science and spiritual endeavor brought into a living parallel. There are other examples as well, and they can ground our attempt to reenvision the relationship between science and religion. In chapter 7 we will revisit Pauli and Jung's work. For now, their story reveals the simple fact that many scientists have spiritual experiences, and those experiences are often filtered through their work as investigators. In Pauli's case the dreams that called him to a deeper exploration of his interior life were rife with the imagery and symbols of theoretical physics. Jung was so impressed by these dreams that they became source material for his own research.[4] At the same time it was through his work with Jung that Pauli gained awareness of the sacred character of his own experiences within physics and his life.

Because the paths linking science and the experience of that world as sacred have been largely lost to us, we now pay a visit to four scientists. Through the varieties of their religious experiences we can put flesh on the ideas explored in the previous chapter.

SCIENTISTS AND RELIGION:
VARIETIES OF EXPERIENCE

Varieties I: Ursula Goodenough

What is being religious anyhow? What about the way I
feel when I think about how cells work or creatures evolve?
Doesn't that feel the same as when I'm listening to the
St. Matthew Passion or standing in the nave of the Notre
Dame Cathedral?

Ursula Goodenough,
The Sacred Depth of Nature

Is Ursula Goodenough a special case? A biologist with an international
reputation and a best-selling textbook on genetics, she has written
widely on science and religion. With great eloquence she has articulated science's capacity to evoke feelings she has experienced in spiritual practice. She is liberal in her interpretation of religious sources
such as the Bible but is clear and determined in her understanding that
human life has an irreducible spiritual dimension. In considering science and religion, she has coined the term *religious naturalism.* Instead
of dwelling on the nature of a supernatural deity, her concern focuses
on scientific narratives of evolution, both living and nonliving, and
their capacity to express the great mystery that underlies all human
experience. Goodenough tries to give the imperatives of spiritual aspiration an expression in scientific understanding. In her own words, "A
religious naturalist is anchored in and dwells within her understandings
of the natural world. He finds his primary religious orientation within
that narrative/perspective and develops mindful religious responses to
it — interpretive, spiritual, and moral."[5]

Goodenough is not the only scientist who has such feelings, but her
history is telling and perhaps gives her a unique capacity for holding
such sensitive and nuanced views.

Goodenough's father, Erwin, was an influential scholar in the history of religion. His work on the symbols of Jewish religious practice

is still held in high regard. Many would hold that he anticipated Joseph Campbell's insistence that myth plays a central role in the development and experience of spirituality in human life.[6] Thus Ursula Goodenough grew up in a household where religious feeling was treated as a living reality. Regardless of his own religious interests, Erwin felt no need to constrain those of his daughter. After finding that her calling lay in science, the former minister and Yale professor exclaimed, "Ursula a scientist! How splendid!"[7]

Many years later, after decades of playing it "straight" as a scientist and highly respected professor herself, Ursula Goodenough found she kept returning to her father's questions. "Why are people religious?" she asked herself. "Why am I not religious?" These questions led her back to an exploration of her own spiritual experience and an attempt to find the meaning of her parallel experience as a scientist.

"So I joined Trinity Presbyterian Church," she writes, "and have spent the last twelve years singing in the choir, reciting the liturgy and prayers, hearing the sermons, participating in the ritual. I came to understand how this tradition, as played out in a middle-class white congregation, is able to elicit states of serious reflection, reverence, gratitude and penance." But even as her own experience deepened she remained ill at ease. There was a dissonance between the context of her religious life and her life as a scientist. "All of it was happening," she writes, "in the context of ancient premises and a deep belief in the supernatural." It was the emphasis on the supernatural that was a thorn in her scientist's paw. *"What about the natural?"* she asked herself. "Was it possible to feel such religious emotions in the context of a fully modern, up to the minute understanding of nature?"[8]

It was in response to these questions that Goodenough wrote her extraordinary book *The Sacred Depths of Nature*.[9] This small volume recounts the "fully modern understanding of nature" in chapters beginning with the Big Bang and continuing through the evolution of life. At the end of each chapter she reflects on the narrative just recounted. Goodenough shows how each scientific story can serve as a gateway to the serious reflection she finds is the hallmark of spiritual or religious feeling. Experience, this time the experience of science's vision of the cosmos, is once again the key. In *The Sacred Depths of Nature* Goodenough points us in a direction rife with possibilities.

So is Ursula Goodenough a special case? While her blending of scientific aspiration and sacred apprehension is original, Goodenough stands in a long line of scientists who have encountered science as an expression of spiritual experience. Many of these investigators had little difficulty balancing the perspectives offered by each endeavor. Some were iconoclasts in their spiritual beliefs about God, and some were traditionalists. Some, like Isaac Newton, were both.

Varieties II: Isaac Newton, Theologist

Isaac Newton, born in 1642 (the year Galileo died), is rightly considered by many scholars the greatest scientist in history. The full extent of Newton's genius was apparent early in life. In his college years at Cambridge University, Newton invented calculus, articulated the nature of light, and developed the laws of gravitation and the general principles of mechanics for which he is now famous. His revolutionary works *Principia* and *Opticks* stand as great (if not the greatest) monuments in scientific thinking alongside Copernicus's *De Revolutionibus* and Darwin's *Origin of the Species*. Alexander Pope's famous epitaph exemplifies the esteem Newton commands,

> Nature and Nature's laws lay hid in Night:
> God said, *Let Newton be!* and All was Light.

Newton has been popularly envisioned as the model of Enlightenment man. For scientists he has been an example of what scientific genius can offer, leading humanity to an age when reason would be the sole guide in revealing the world's great truths. But Newton, it turns out, was far too complex a personality to pigeon-holed. Recent scholarship has made clear the full extent that religion played in his life.

By all accounts Newton was a strange and difficult man. In spite of garnering considerable fame and fortune, he remained insecure and subject to bouts of intense rage and depression. Though he was capable of strong friendships, he never married and is said to have remained a virgin all his life. Newton took on a variety of roles during his lifetime. In 1696 he became warden of the Royal Mint, and in 1703 he was elected president of the Royal Society, the premier scientific organization of its age. Until recently, however, one aspect of Newton's life, the religious, has remained in the shadows. Given his status as a hero of the age of reason, many in the sciences and humanities were loath to deal with the extent of Newton's religious convictions. He spent many years engaged in theological study, perhaps more time than he devoted to science. His topics were hardly the kinds that a modern scientist would find enlightening: the nature of the Trinity (a concept he vehemently opposed), alchemy, and the establishment of a more correct version of biblical prophesy.

Newton did not see a conflict between his scientific thinking and his religious convictions.[10] Instead he fashioned himself a priest of nature. The harmony he found in his science was evidence of God's greatness. As he wrote in one of his theological manuscripts, "It is the perfection of God's works that they are all done with the greatest simplicity. He is the God of order & not confusion." Throughout his life Newton took his work to be that of reformer in the domains of both natural philosophy and religion. The two were, for him, inseparable. "There is no way (without revelation)," he wrote, "to come to the knowledge of a Deity but by the frame of nature."

Newton was passionate in his belief that there existed a God of absolute power who can and did intervene directly in the human and natural worlds. It is notable that Newton carried out his theological studies in private, refraining from allowing them to be published. This may simply have been a matter of self-protection. Denial of the Trinity was a punishable offense. It may also have simply been snobbery. Newton felt that his views on religion were too sophisticated for the common man.

Whatever his reasons, modern scholarship shows Isaac Newton as a man of profound spiritual sensibilities. It is a great irony that there was no warfare between science and religion for the world's greatest scientist.

Varieties III: James Clerk Maxwell and the Glory of God

> I think that Christians whose minds are scientific are bound to study science that this view of the glory of God may be as extensive as their being is capable of.
> James Clerk Maxwell, quoted in Raymond Seeger,
> "Maxwell Devout Inquirer"

A lifelong concern with the God of the Bible also characterized the great physicist James Clerk Maxwell.[11] He was born in Scotland in 1831, blessed by wealth, an even temper, and a disposition that easily won him friends. Despite his early death, at the age of thirty-eight, Maxwell's contributions to physics were profound and varied. His greatest work was the formulation of equations that govern electric and magnetic fields and the unification of these phenomena into electromagnetism. The beauty and power of the four simple, elegant mathematical laws, known as Maxwell's equations, are such that they describe *all* the possible behaviors of macroscopic electric and magnetic fields. It was Maxwell who predicted that light could be described as moving waves of an electromagnetic field. We can thank Maxwell for radio, television, the cell phone, and, in part, just about every other technological marvel that involves electricity and magnetism. Before Maxwell it was unclear how a cause occurring at one place could lead to effect someplace else. Physicists had, for example, long been bothered by the strange "action-at-a-distance" that happened with Newtonian gravity. How did the Sun reach across empty space and exert an invisible gravitational tug on the Earth a hundred million kilometers away? This spooky action-at-a-distance could be done away with by invoking a gravitational field that extended throughout space. The concept of "fields" began with Maxwell and his study of electromagnetism. It was a fundamentally new idea, and it was driven into physics through his genius. Fields now form the basis of essentially all modern physical theory, demonstrating the extent of Maxwell's influence.

Maxwell was also a devout Christian. In his college years he wrote to his father quoting the famous dictum, "I believe . . . that Man's chief end is to glorify God and to enjoy him for ever." He remained a rigorous Presbyterian throughout his life, reading Scripture each night with his wife, Katharine. He was an elder in his church and was dedicated enough to commit the entirety of the psalms to memory. His belief, it appears, never wavered.

As his fame grew he was called upon to address issues in the overlap between science and religion. A century and a half after Newton, the primacy of science was solidifying. In the wake of the Enlightenment, discussion about the domains of science and religion had changed. Promethean science was on the ascent. It impinged on religion's claims of revelation about the natural world. Religious scientists such as Maxwell found themselves straddling spiritual beliefs and an understanding of nature gained through scientific investigation. Maxwell appeared to rest easy in his balance of the two worlds. In a letter to a local vicar Maxwell cautioned the man against using science to attempt to firm up Scripture:

> But I should be very sorry if an interpretation founded on a most conjectural scientific hypothesis were to get fastened to the text in Genesis. . . . The rate of change of scientific hypothesis is naturally much more rapid than that of Biblical interpretations, so that if an interpretation is founded on such an hypothesis, it may help to keep the hypothesis above ground long after it ought to be buried and forgotten. At the same time I think that each individual man should do all he can to impress his own mind with the extent, the order, and the unity of the universe, and should carry these ideas with him as he reads [the Bible].[12]

In the last sentence of this passage one can see the interweaving of scientific and religious inspirations. The profound order inherent in the physical universe enriches spiritual understanding. The experience of the world through the lens of spiritual experience makes science a sacred activity. Maxwell spent his working life focused on theoretical physics alone. Unlike Newton, he did not devote himself to theology. Still, his personal letters show that his experience of science was one aspect of a deep and abiding aspiration to know and glorify his God.

· · ·

Maxwell, along with Newton, stands as one pole in the religious expression of individual scientists. They are, in a sense, traditionalists (though Newton was at odds with the religious orthodoxy of his day). They have no quarrel with a spiritual life that takes the personal God of the Judeo-Christian tradition as its focus. In 1997 a *Scientific American* poll of scientists found that 40 percent believed in a personal god (this number was remarkably unchanged from a similar poll taken in 1920). These scientists would, apparently, align themselves with Newton and Maxwell.

From the Enlightenment onward, however, many scientists began to find themselves in the remaining 60 percent. For these scientists a God who directly intervenes in the world stands in direct opposition to science's efficacy and power. For those who reject such a notion of divinity and yet still respond to the world through religious experience, new descriptions would have to be found. These would often be idiosyncratic as each researcher, like Ursula Goodenough, struggled to balance the call of the sacred with his or her integrity as a scientist. Somehow this sense of the world as sacred would have to be balanced with a rejection of supernatural causes. This form of scientific religious expression was most famously embodied in Albert Einstein.

Varieties IV: Albert Einstein and Scientific Spirituality

Albert Einstein is the most universally recognized face of science on the planet. The images of Einstein in his later years, grandfatherly and with an electric mane of white hair, careening around Santa Barbara on a bike or sticking his tongue out at the camera, have come to symbolize both the wisdom and the playful creativity that people hope for in their geniuses. Scientifically, it is almost impossible to overstate his importance. Einstein played an instrumental role in both great scientific revolutions of the twentieth century. His theory of relativity overthrew the conceptual underpinnings of Newton's physics, initiating a fundamentally original understanding of space, time, matter, and energy. His early work on the "photo-voltaic effect" — electric currents formed when light shines on a conducting surface — helped lay the groundwork for the emerging quantum revolution. As his fame from these achievements grew, his stature as an international figure did as well; and he became a relentless advocate for peace and social justice.[13]

Because of this stature Einstein's views on religion have often been quoted, or misquoted, to bolster positive views of a concordance between science and religion. A deeper understanding of Einstein's work and philosophy, however, shows that one must be careful to place his religion in the proper context. Einstein's sense of spirituality was wrapped around a firm belief in the determinist nature of the universe. If he had a faith it was a faith that the universe was comprehensible through unchanging physical laws that ruled all aspects of existence. Einstein's very definite spiritual sensibilities grew from soil watered by mathematical physics.

Though a lifelong supporter of Judaism, he left the faith of his fathers early on. He wrote of his early life, "I came . . . to a deep religiousness, which, however, reached an abrupt end at the age of twelve."[14] This abandonment of religion was linked to his distaste for traditional concepts of God. Unlike Newton or Maxwell, Einstein could not bear the idea of a personal god. When asked by a rabbi about his beliefs Einstein replied, "It seems to me that the idea of a personal God is an anthropomorphic concept which I cannot take seriously."[15] Later he wrote, "I do not believe in the God of theology who rewards good and punishes evil. My God created laws that take care of that. His universe is not ruled by wishful thinking, but by immutable laws."[16] Einstein, a rationalist throughout his life, remained deeply suspicious of institutionalized religion and "the formation of a special priestly caste which sets itself up as a mediator between the people and the beings they fear, and erects hegemony on this basis."[17] Taken from this perspective those who cite Einstein to bolster the case for the commensurability of science and religion must do so carefully or at their peril.

In spite of his unorthodox views Einstein saw himself as a man of deep spiritual sensibilities. He had no patience with atheism and could not understand "the fanatical atheists whose intolerance is of the same kind as the intolerance of the religious fanatics and comes from the same source. They are creatures who — in their grudge against the traditional 'opium of the people' — cannot bear the music of the spheres."[18] For Einstein the universe itself was an enigma at its very core, and this was the first cause of religious feeling. His spiritual sensibilities emerged from his experience of the world through his beloved physics, the music of the spheres.

The experience of mystery was for Einstein the experience of the spiritual. "The fairest thing we can experience is the mysterious,"[19] he said. "It is the fundamental emotion which stands at the cradle of true art and true science. He who knows it not and can no longer wonder, no longer feel amazement, is as good as dead, a snuffed-out candle." This mystery is, for Einstein, the source of the scientist's inspiration. "You will hardly find one among the profounder sort of scientific minds without a peculiar religious feeling of his own. . . . But it is different from the religion of the naive man. His religious feeling takes the form of a rapturous amazement at the harmony of natural law, which reveals an intelligence of such superiority that, compared with it, all the systematic thinking and acting of human beings is an utterly insignificant reflection."[20]

The intelligence Einstein speaks of would, however, be no balm to the Intelligent Design enthusiast. Einstein's conception of God and divinity was drawn from his readings of the seventeenth-century Dutch

philosopher Baruch Spinoza. Spinoza has been called a natural pan-
theist because he identified God directly with nature and nature with
the underlying substratum of deterministic laws. In this sense there is
nothing supernatural about the source of religion. In a sense God *is*
physical law and is not above nature. Thus Einstein's religious feeling
differs appreciably from that of Newton and Maxwell. I suspect that
many modern scientists would be far more comfortable with Einstein's
approach than with that of his English predecessors.

ODIN AT THE WELL OF MIMIR:
THE PRICE OF WISDOM

Bitter winds blew in from the dark mountains of Jotunhiem, the terrible realm of the giants. Odin drew his blue cloak against the wind and
stopped to survey his progress. It had been days since he left Asgard,
city of the gods, and crossed the rainbow bridge into Midgard, the
realm of men. Beneath the sting of the icy gale he could still feel the
gentle touch of his beautiful wife Frigga's kiss as he stood at the gate
between the two worlds. "I know the signs and portents weigh on you,
husband," she whispered. "I know you make for Mimir and his well."
He could not give answer then to the fear in her eyes. "Tomorrow,"
he said, "tomorrow I will not be Odin but Vegam, wanderer upon the
ways of Mitgard and Jotunhiem." Without looking back to her, he had
exchanged his golden armor and great spear for the simple blue cloak
and stepped across the threshold.

Now he regretted not being more direct with his wife. Her love had
always sustained him. Even Odin All-Father, mightiest of the gods,
knew fear. If he trembled at what waited before him, what must his
loving companion have felt? "I do what must be done," he reminded
himself, "for the good of the world. I travel to turn what knowledge I
have into wisdom."

The wars between the gods, who would make the world better for
men, and the giants, who would throw all to cold and darkness, had

waged for eons. Only days ago Odin's emissaries had told him of the great evil that hung over the Future in the wake of these wars. But all his emissaries could give him was a warning, incomplete and troubling. He still did not know what form this great evil would take or its appointed time. Odin set out on this journey to change that. He had left Asgard to lift his ignorance. His questions could be answered if only he was willing to pay the terrible price, the price none before him had been willing to pay. After a moment's more reflection Odin pushed away his fear and quickened his pace on the path to Mimir's well.

Mimir, guardian of the well of knowledge, was gnarled and bent, but behind his hooded eyes there was power. For uncounted years before the dwellers of Asgard built their great city, Mimir dwelt here guarding these sacred waters. Each day he dipped his horn into the well and drank deeply. All the world's knowledge and wisdom were his. Thus when the traveler in the blue cloak arrived Mimir knew immediately who stood before him.

"Hail, Odin," said Mimir, "eldest of the gods."

Odin made his reverence to the ancient and most wise of the world's beings. "I would drink from your well, Mimir," he said.

"Are you prepared, All-Father to the gods, to pay the price?" asked Mimir. "The price all before you have shrunk away from? Will you give up your right eye to drink from my well?"

Odin too almost shrank from his quest. To go through the eons with only one eye, half sighted before the world's impossible beauty was too much to ask. But he remembered what was at stake and what hung in the balance. "I will pay your price," said Odin in resignation.

"Then drink, eldest of the gods."

Odin drew the horn through the water and drank the sweet waters. He drank and all the world's knowledge and all its future was laid out before him. He saw Time unfurl like a dragon's tongue. He saw the Universe and its ways. He saw the stars spinning in the black with their risings and fallings. At the end of it all Odin saw the Twilight of the gods. Laid out before him in his mind's eye he saw the great battle between the Gods and the Giants that would end the world. He saw the suffering and torment of men as all Creation shuddered and collapsed. And then, amid this vision, he saw why it must be so. He saw the play of time. He saw what he might do to ensure that, even in the gods' destruction, Evil would be defeated and the seeds of a better world would be sown. All this he saw as his drank from the well of Mimir.

When the vision had passed he laid down the horn. Then Odin, the All-Father, put his hand to his face and tore out his right eye. Anguish and pain the great god knew then. He sank to his knees and wailed as if a great fire were consuming him. Mimir, paying no heed, took the eye of Odin and dropped it into the well, where it remains to this day as the price paid for the world's wisdom.

KNOWLEDGE, WISDOM, AND EXAMPLE

> A bit beyond perception's reach
> I sometimes believe I see
> that Life is two locked boxes, each
> containing the other's key.
> Piet Hein,
> "Grooks by Piet Hein" Web site

We all have to pay a price for knowledge and understanding. In the myth of Mimir's Well, Odin had to give up a cherished part of himself to understand how the world was structured. In the debate between science and religion we will have to give up our stereotypes and the desire for simple answers to gain an understanding of the world's most subtle ways. No simple formula cast from a hundred retellings of history wrongly recalled can serve our needs. No easy reliance on someone's scriptures, or the rejection of those scriptures, can teach us the language of heart and mind as they burn in the constant fire. No cartoons or caricatures will work here, and we must abandon them.

Newton, Maxwell, Einstein, and Goodenough provide four very

different perspectives on the way scientists can simultaneously pursue their investigations and maintain a sense of the sacred necessary for religious feeling. Thus on the level of individual scientists there is no question that religion and science can co-exist. The problem for us is that each individual has had to work out, for himself or herself, the meaning and extent of commingled spiritual and scientific sensibilities. This is not of much help to a society in which conflict between science and religion becomes a cultural standard.

We have already seen that there is a different approach. The emphasis on what James calls religious experience takes us a long way toward shifting the center of gravity in the discussion of science and religion. The four examples explored in this chapter make it clear that a more fruitful direction exists. So many scientists have their own versions of spiritual experience brought to them through their work. Why should we continue the old debate that compares empirical results with religious metaphysics? Instead we can see science and the vision of the universe it offers as a series of hierophanies. Science becomes the location where the world can be illuminated as sacred, as worthy of awe and reverence in and of itself. That perspective would amount to a radical departure from a century or more of contention. Seeing science as a means to manifest the quality of experience we call sacred is a first, critical step toward developing a fundamentally different way of understanding these two great human activities.

But we can, and must, go further still.

Science is not simply an invention of the past five centuries. It did not spring fully formed from the minds of Copernicus or Newton or Darwin. If science is a means to evoke the sacred it connects us to an endeavor that traces its roots back through millennia, touching the great Greek philosophers and the builders of Newgrange in Ireland. If science creates hierophanies it is part of an aspiration that has been with us since our origins. Our inability to recognize these functions in science comes because we cannot see that it and religion are part of a much older impulse. We must now look back to see where the roots of hierophany emerged. Long ago humanity stood watchful for the ways the world made itself known. This readiness is encoded in our oldest attempts to make sense of the world and meet the sacred. We now take our next step toward a renewed perspective on science and religion, and that step goes far backward. Only by looking back to our origins can we see the common root of both our spiritual and scientific impulses in the ancient universe of myth.

CHAPTER 5

Science, Myth, and Sacred Narratives

The Universe as Story

I am among those who think that science has great beauty.
A scientist in his laboratory is not only a technician: he is
also a child placed before natural phenomena which impress
him like a fairy tale.

Marie Curie, quoted in Eve Curie,
Madame Curie: A Biography

Myth embodies the nearest approach to absolute truth that
can be expressed in words.

Ananda Coomaraswamy, *Hinduism and Buddhism*

Truth is like a fire and will burn through and be seen.

Maxwell Anderson, *Winterset: A Play in 3 Acts*

The third-graders sit in a constantly shifting half circle. In spite of
their teacher's protests, they squirm and bounce, blurting out astronomy facts and non sequiturs. The kid with the blue shirt and runny
nose knows the speed of light to three significant digits. The girl with

the checkered pants has a hamster named Biz who got steamed to death over a heat vent last night.

It's ten in the morning at the Rush Elementary School, twenty miles outside of Rochester, and I'm hoping we're all having fun. I'm here attending a question-and-answer session with students who just finished the astronomy section of their earth science curriculum. A room full of eight-year-olds makes for an enthusiastic crowd. They have that particular kind of wonder that only dinosaurs and astronomy can produce in kids. Questions ranging from the origin of comets to why everything in space spins are thrown at me. Eventually, one of the kids asks me if the Sun will blow up. I always get the blowing-up question. Sometimes it's about the Sun, sometimes the Earth. Once, a boy asked me if he was going to blow up.

I explain that five billion years in the future, our Sun will run out of fusion fuel and swell into a red giant. Everything out to Earth's orbit will be swallowed by the bloated dying star. Most of the students think this is incredibly cool. They clamor for more details about the world's fiery end, which I dutifully provide. The questions are the best part of my work with kids. It's all so new and fresh and *big* that they can barely contain themselves. I wish my freshman classes showed such enthusiasm.

Soon the time comes to finish up. I say my good-byes and gather up my stuff: the laptop with Hubble images, the model of the Earth, and the small telescope. As I slip on my coat the class is making its own noisy preparations. From the side a little girl approaches me in tears. I ask her what's wrong. Through snuffles and nose-wiping on her sleeve she tells me the end of the world scares her. Now I feel really bad. I tell her that she doesn't need to be scared because the Sun will live for another five billion years, which is a really, really long time. "Everybody will have found a new home by then," I say. Not surprisingly, this abstraction provides no consolation to her. I try a different tack. "You know, it's so far away in the future that you could think of it like a real-life fairy tale." We talk about this idea for a moment, and then she asks, "So it's true, but it's still just a story?" I tell her that's about right. She wipes her nose on her sleeve again and says a bit more brightly, "I guess that's okay." Then she turns and walks back to her class. Only later, in Newgrange, did I realize how right she was.

. . .

What were we doing all those years? What were we thinking, dreaming, discovering through those countless days and nights stretching across millennia? Human history began some five thousand years ago. But human evolution spans a time scale dwarfing that history. Our first ancestors appeared almost four million years ago. It took less time for the Sun to form from its natal gas cloud (a mere million years) than it took for humans to evolve from the first bipedal primates.

Our ancestors awakened slowly. They created culture in the clothes they fashioned and the way they buried the dead. Tens of thousands of years before we stepped into written history, they created richly imagined paintings on the walls of caves and escarpments of rock. In dreamlike images of animals, hunters, and pregnant women they translated their experience into symbols and form. These first works of art expressed their vision of the world around them and the life within.

We have been watching, acting, and dreaming for a very long time. If experience forms the impulse of what we now call religion it was present for our forebears too. They watched closely and felt deeply. In this way we are no different from them. The impulse that became science was also with them. Through need and curiosity, they saw the patterns in the world. They saw the rising and setting of Sun and Moon, the returning of season and storm. The constant fire, the response to the world's beauty and power, was with them as it is with us. It was this original response, this first aspiration, that grew into science and religion. It has never left us. This constant fire remains the essence of our response to the sacred in experience. It has always been with us, captured in the universe of myth.

Myth is the space where humanity imaginatively represents the world and its discovered truth through sacred narratives. The importance of myth is its ability to transmit essential truths with poetic economy. Myth, in this essential context, does not mean "false story." It is far more essential and far more powerful. Every culture has its mythology. Every culture has its potent narratives of origins and endings. If we wish to follow the harmonizing roots of science and religion, then we must turn to the universe of myth. Within it we will find an original response to the one mystery of Being that all of us have faced in the long march of time.

If science and spiritual endeavor have a connection that transcends the comparisons of results, then it must be rooted in something that precedes them both. That is why we turn now to myth. Myth came before all our forms of religion and before the practice of science. In myth the experience of the world as sacred was codified in stories that contained the origins and aspiration of both science and religion. By understanding myth, its forms and functions, we can complete our trek. It brings us to the vantage point that reveals a new and more enlivened vision of science and spiritual endeavor.

The hierophanies particular to science occur within science's own narratives. Science tells us stories: how stars are born in the immensity of space; how thunder and lightning can tear the very air apart; how human beings evolved across unmeasured seasons. These stories can also be seen as sacred narratives. Through them we experience the vertigo of awe and the weight of the world's mystery pressing down on us. Through them we are called to an experience that sweeps the profane, the unrecollected, away. The impact and imperative of these narratives are not different in kind from the ones our ancestors heard. It is the all-important aspiration to the True and the Real that remains unchanged. In that aspiration we can find the authentic connections between science and the domains of spiritual endeavor. As the astronomer Edwin Krupp once wrote, "We aren't really more mature than our ancestors were but we do know a few things they did not."[1]

In this chapter we will consider science in light of mythology and our enduring compulsion to make sense of the world through story. The interest in myth and mythology grew throughout the twentieth century. It began first among scholars, then, thanks in large part to Joseph Campbell, in the popular imagination. We will consider a variety of voices: scholars and writers who have shed light on the purpose of these ancient stories. In this light we will see science and religion as an ancient but continuing response to the experience of the world's mystery and order. In that response, in science seen through the lens of myth, we may also find the capacity for a renewed response to the overwhelming challenges we face today.

The origins of myth are lost across the horizon of our cultural memories and the artifacts of our evolution. To understand what myth is and what it means for us today we have to begin with a story: the science of human origins. Myth emerged from our own long evolution. From protohominids to the genus *Homo* we have adapted and transformed. At some point in the dim past we became self-conscious. That is where myth begins. The story of our awakening into mythmakers, of our awakening to the world made sacred, is our starting point.

A HISTORY BEFORE HISTORY: HUMAN ORIGINS

We are the sole survivors. Our species, *Homo sapiens*, is the last one left in the genus *Homo*. All the others, our cousins in the long march of evolution, have left the stage. In their extinction, the gift of this verdant blue-green world was bequeathed solely to us.

Way back in high school biology we learned the basic levels of biological classification: Domain; Kingdom; Phylum; Class; Order; Genus; Species. This scheme groups all creatures, living and extinct, based on their physical characteristics. The classification makes a tree of life, with Domain forming the trunk and Species forming the branches. Modern humans belong to the order Primates. We share this order with lemurs, tarsiers, Old World monkeys, New World monkeys, and apes. Primates are distinguished by forward-looking eyes (providing 3-D vision), hands built for fine motor skills, and enlarged brains relative to body size. Primate evolution can be traced to a common ancestor more than 55 million years ago. From there evolution carried the primates forward to greater division. At some point between 8 and 5 million years ago the line linking chimpanzees, gorillas, and humans split apart.[2]

The drama of our evolution began a few million years later. The first uniquely human ancestors were bipedal creatures that struggled to draw themselves upright.[3] The evidence for these early walkers is remarkable. In the dry flat plane of Laetoli, Tanzania, paleontologists have found ancient footprints preserved in hardened volcanic ash. The impressions were laid down 3.5 million years ago. These footprints are the silent testimony of *Australopithecus afarensis*, a creature that moved on two legs across a world of broad savannah and lush woodlands. From its footfalls onward scientists can trace the outline of our subsequent development in a story of structural changes in skull, jaw, pelvis, and other body forms. Many of these changes are likely mirrored in the growing development of intelligence. As the eons pass the story becomes one of growing sophistication in the application of that intelligence: the use of tools and the development of culture.

Our use of instruments to aid the work of survival began early on. By 2.5 million years ago the creatures we call *Homo habilis* were already fashioning stone tools in the form of hammers and scrapers. One million years later *Homo erectus* was using hand axes for butchering and woodworking. Our mastery of the environment grew and soon included the use of fire, a critical step in the evolution of intelligence. Charred animal bones found at sites such as the Swartkrans region of South Africa suggest that more than one million years ago our ancestors had tamed fire for their own uses.

The toolmaking creatures that led to modern humans remained in this African home for three or four million years. Slowly, our ancestors radiated outward into Europe and Asia. They took their inventiveness and growing technical capabilities with them. By 150,000 years ago the Neanderthal and other human cousins occupied locations as far-flung as Croatia and China. Everywhere they lived our ancestors refined their methods for creating stone tools. Their blades became more uniform and more plentiful. Between 40,000 and 16,000 years ago *Homo sapiens'* penchant for toolmaking exploded as simple blades turned into fishhooks, needles, and harpoons. Understanding how to control fire also continued; as early as 100,000 years ago hearths became common in human camps and caves.

This story of toolmaking and the domestication of fire is a narrative of creatures becoming increasingly sophisticated, manipulating both objects and processes to ensure survival. It is also a story of self-consciousness dawning. What emerged in our ancestors during this time was an interior life that responded to the world as it was encountered. Death, the great mystery, appears to bring some of the first and most articulate responses. Evidence suggests that by 70,000 years ago Neanderthals began burying their dead. At numerous sites around the globe paleoanthropologists have excavated bodies placed in special "flexed," or folded, positions. Artifacts such as spear points were placed in these graves alongside the bodies.

The importance of these first burials cannot be overstated. The care taken with the dead indicates that early humans began to apprehend death as a transition. The rituals of burial may have been intended to mediate that transition by honoring the dead and preparing them for a life to come. They speak to an awakening of human beings to their own condition. This flowering of consciousness is also given voice in a second, equally crucial revolution — the appearance of art. In Australia petroglyphs, along with zigzag lines and dots incised in rocks, have been discovered that date back 40,000 years. In the Divje Babe Cave in Slovenia a 30,000-year-old bone flute with four holes (producing a diatonic scale) appears along with other artifacts of primitive habitation.[4] Cave paintings of animals appeared in a variety of settings around 30,000 years ago.[5] Within 10,000 years human figures were included along with the animals. This change testifies to our emergence as players in our own story. Soon afterward "Venus" figures with exaggerated sexual characteristics began to appear. These statuettes, fashioned by forgotten artists around the sparsely populated planet, show humans transforming their experience into representation and symbol. In many cases these early expressions of artistic consciousness are found in places so difficult to reach that they must have been created in special ceremonies or rituals. On the desolate Nullarbor plain in southern Australia, for example, a remote cave system known as Koonalda reveals walls covered with zigzag incisions cut by human hands 20,000 years ago. The incised walls are 200 feet below the surface, difficult to reach even today. The famous cave paintings at Lascaux in France, with their rich imagery of animals and human handprints, can only be entered after a long, slow climb through narrow passageways. Something special happened in these places. One can imagine these early artists making arduous journeys to their sacred caves because some internal illumination was thrown upon experience there. Then in the process of creation some power was, perhaps, gained.

The burial sites and early artworks give silent testimony that far back in prehistory our ancestors attempted to fashion some understanding of their experience. As Karen Armstrong has written, "The Neanderthals who buried companions with such care . . . imagined that the visible material world was not the only reality."[6] These activities were likely accompanied by stories that illuminated their meaning and set them in context. Myth began during these uncounted eons and evolved along with us.

A STORY BEFORE TEXTS: THE HISTORY OF MYTH

We started as small tribes of hunter-gatherers. In time this nomadic life, moving with the animals we hunted and the food we could find, gave way to agriculture. The methods of sowing and harvest created surplus, and we learned to build cities. As the cities grew their societies became more complex. The art of writing was mastered, and we emerged into our own history. All that time, across all those millennia, we told stories. That was a constant. As we changed, the stories changed. The shifts in narrative reflected both our growing understanding and our changing psychological needs. What did not change was the function of these stories, these great myths. Their narratives gave our collective life its meaning. The myths told us what this world was and how it came to be. They told us who we were and how we should behave. Always myths spoke of the deepest encounters with our own sacred depths, the unseen potential manifested in our hearts and experience.

In *A Short History of Myth*, Karen Armstrong draws diverse sources together to recount forty millennia of human mythmaking. Drawing heavily on Eliade, Armstrong reminds us that myths are narratives that speak of what lies beyond or below the visible world. They are always true, but they are more than simply right. "A myth," she writes, "was an event which, in some sense, had happened once but which also happened all the time."[7] Myths begin with what is before us and then transform it into a narrative that transcends the profane time of everyday life. Armstrong's account traces the history of myth as a story of our cultural evolution. It tells us of changing encounters with the world that our evolution made possible.

The Paleolithic Era

The mythology of the period between 20,000 and 8,000 years ago comes to us in fragments. Some stories survived in the literature of later cultures. Some are found in the living myths of indigenous peoples such as the Australian Aborigines. From the work of archaeologists, ethnologists, and anthropologists comes a vision of the Paleolithic cosmos as told through its mythology. It was a universe rich in domains of the spirit, a universe in which all aspects of life were imbued with the luminosity of the sacred. The world was alive, and humans lived in close proximity to the source of its animate powers.

One of the earliest, most prevalent myths spoke to its people of a previous golden age. As Armstrong writes, "The spiritual world was such a compelling and immediate reality that . . . it must have once been more accessible to human beings."[8] There was a lost paradise somewhere not too far off in time and space. There humans knew harmony. In some myths human and animal could speak each other's language. This harmony was broken somehow, but people could reestablish and maintain close contact with the sacred through the telling of the myths and the enactment of the rituals that accompanied them.

Another common type of myth of the Paleolithic era is the journey of ascent. These are myths of transcendence in which a hero makes a journey to a hidden realm for the good of his people. Along the way he suffers great loss but is also transformed, becoming more than his former self. The origins of this common theme appear to lie in the danger and violence of the hunt. The difficult practice of hunting continually demonstrated the painful reality of humanity's loss of a first, golden age. In the wake of those experiences a mythic response was required. The chief prey of many Paleolithic tribes was large mammals. Their cries, facial expressions, and flowing blood were familiar enough to humans to elicit a sympathetic response. The journey of ascent grew with other myths to relieve the anxiety that sympathy evoked and to place the hunt in a wider framework. Thus the hunters' rites of initiation flowed together with myths of the hero's dangerous journey. The link between the physical and spiritual worlds was established by the shaman, who was spiritual leader of the tribe. Shamans made their own journeys. While hunters undertook their dangerous treks for the good of the tribe, shamans in trance states traveled to the realms of the gods when conditions such as sickness or loss of game demanded it. In both cases transcendence and ascent served to heal the loss of the golden age and allow humans to rise above the dangers so prevalent in their daily lives.

Armstrong identifies twin responses to the world emerging at this time. *Mythos* is the felt reaction, the interior resonance that animates metaphor and symbol. *Logos* is the rational, studied response. It is the planning and consideration that leads to advances in technology. Like the sacred and the profane, during this time Mythos and Logos were intertwined.[9] Feeling and thinking, spirit and object were still of a single piece. Thus, while the myths speak of the psychological and spiritual needs of early humans, the close observation of the world is there as well. After hunting expeditions the skeleton and pelt of an animal would be laid out in ritual to reconstruct it and give it new life.[10] Just as today, early humans could watch and note how the world around them appeared to be constructed. They could try to use what they found to understand and control the world. Hunting required close cooperation among expedition members and increasing sophistication in toolmaking. Close observation was also required to know which plants to gather and when. Thus Logos, required to survive, became part of myth and ritual.

The close observation of the world also appears in the myths of the sky. These are perhaps the earliest and most universal mythic narratives told. The first conception of the divine came from the experience of the immensity of the night sky as inaccessible and eternal. The sky was both remote and dynamic. It was a ceaseless play of storms, thunder, rainbows, eclipses, comets, and meteors. All this activity played out against the regular movements of Sun, Moon, and planets. The sky was the original hierophany: a gateway into and manifestation of the sacred. As human culture evolved our response to the sky did as well. At some point the sky became a god.

Armstrong describes a turning point late in the Paleolithic era when the sky became personified. This shift marked a profound transformation in the development of human understanding. For Armstrong, myths make the sacred manifest when they are experienced as close at hand. The conception of a sky god signified a retreat of this power into absence. In the myths of the late Paleolithic era the sky god became all-powerful but often had no dealings with the world of men. According to Armstrong, the change of the sky into a sky god showed how the world of humans was changing. The retreat of the early sky gods prefigures the radical transformation humanity experienced as the Earth was brought under the weight of the plow.

The Neolithic Era

Humans discovered the science of agriculture almost 10,000 years ago.[11] It was a technological revolution rivaled only by the transition our great-great-great-grandparents faced during the industrial revolution a few hundred years ago. Unlike recent industries fueled by coal and oil, agriculture was a science discovered in closest proximity to myth. The agrarian revolution saw a profound shift in humanity's ability to feed itself and its relation to the sacred. Myths changed along with, and in response to, this reorganization of culture and consciousness.

"Agriculture," writes Armstrong, "was the product of Logos, but unlike the technological revolutions of our own day it was not regarded as a purely secular enterprise."[12] Farming became as much of a sacred act as hunting was for the Paleolithic humans. Crops, emerging from careful seeding and cultivation, were a hierophany. They were a revelation the way a tomato eaten right off the vine can be a revelation to a city kid. Here was a new form of divine power. Unseen energies existed that turned the tilled Earth into a womb for the community.

The rites, rituals, and myths of the Neolithic peoples responded to the needs of agricultural life. The first seeds of each year's sowing were thrown away as offerings. The first fruits were left to drop, recharging the hidden forces animating the agrarian cycle. Ritual sexual union between men and women would precede the planting as a way of mani- festing the hierogamy, the sacred marriage of soil, seed, and rain to achieve fruition. Humans were tapping into vast creative powers. The close observation of the world that made agriculture possible had its internal complement in the transformations of the sacred narratives of mythology.

Awe and wonder before the sky led to sky gods in the Paleolithic era. Now, in Neolithic times, the Earth became a mother goddess. Myths throughout the fertile Middle Eastern cultures tell stories of mother goddesses and their connection to farming. The heroic journeys of the male hunter now gave way to the dangerous travel of mother goddesses descending into the underworld of death and returning to bring new life. These mother goddesses were not passive, purely nurturing creatures, just as the ritualized sex that preceded plantings was not an intimation of romantic love. In the first millennia of farming, agriculture was not a pastoral practice but a struggle. Drought, famine, and the violence of storms and floods were always close at hand. They were reminders of the full potential inherent in the energies of the sacred. Myths expressed the power people felt surrounding them and its devastating possibilities. "Mythology is not escapist," Armstrong reminds us. It arose from experience and in that way was honest. It forced people to face the reality of death and transition.

Examples of the new myths abound in the Near East. The ancient Syrian goddess Anat is one manifestation of the dynamic mother goddess. Anat is the sister of the storm god Baal. After Baal is killed by the god of death, Baal's father, El, can only weep in sorrow. It is Anat who searches for revenge and a way to reanimate Baal. El is an example of the absent and impotent sky god, a remnant of the Paleolithic mythologies. Now it is the goddess Anat who must defeat Mot, the god of death. In perfect symbolism Anat destroys Mot using all the tools of the farmer. Anat fells Mot with a sickle, winnows and grinds him in sieve and mill, and finally scatters his flesh over the Earth. In one version of the myth Anat manages to bring Baal back to life, restoring cosmic harmony through sexual union with him. These narratives of death and rebirth are told many times in many different cultures. In Egypt Osiris, the first king, is cut into many pieces and buried. It is Isis, his sister and wife, who finds the king and revives him long enough to conceive a new regent.

The new myths were sensitive to the cycles of the new agrarian year. In Mesopotamia the goddess Inanna is called upon to make a dangerous journey to the underworld kingdom. In her absence Inanna's shepherd husband usurps her throne, and when she returns he is banished for his crime. The story ends with Inanna's husband spending six months of each year in the underworld. The yearly cycles of life and death also appear in the myth of Demeter and Persephone, which dates to the Neolithic era. Demeter controls the powers of the grain. When Hades, god of the underworld, abducts her daughter, Persephone, Demeter rages and threatens to starve human beings by withholding the harvest. Zeus agrees to rescue Persephone. Once it is learned that Persephone has tasted the fruit of the underworld the return cannot be fully completed and the girl must spend four months of each year with Hades.

These stories may seem to be simple allegories of nature, but they are far more. As Armstrong notes, the rites of Demeter did not correspond with either sowing or reaping. These myths are about both the exterior workings of the world and our interior responses to it. They derive their efficacy from close observation and then go beyond it to touch the core of powers encountered in experience. In these mother goddess stories the core of the story is about death as an essential part of life. While these narratives were animated by the new science of agriculture, they also embody a response to the world appearing as sacred in new ways.

City Building and the Axial Age

By 4000 B.C.E. a new development in human endeavor had taken root. In Egypt and Mesopotamia, and later in both China and India, complex cities arose. Mythology once again adapted. "Human life," Armstrong writes, "was becoming more self-conscious. People could now give permanent expression to their aspirations in the civilized arts, and the invention of writing meant they could give enduring literary expression to their mythology."

For all their glory cities and the civilizations they supported were fragile things. Cycles of building and rebuilding within the fortified walls of a city echoed the dramatic rise and fall of these individual, localized societies. There are many reasons why early civilizations and their cities failed. Sometimes it was caused by invasion. Just as often environmental stresses were the cause. Either way, city building was a tenuous game. And once again the old myths had to give way. New myths arose that reflected an endless struggle between order and chaos, civilization and barbarism.

In Mesopotamia the city itself became the hierophany. Within its walls the sacred was encountered in the temples and through the arts and trades that allowed civilization to flourish. The gods had taught humans to build ziggurats, huge stone structures that replaced mountains as the high place where the sacred manifested. In Babylon the god of wisdom, Enki, oversaw the trades that city building demanded: doctors, potters, scribes, and leatherworkers. People became more self-conscious, more aware of the degree to which they controlled their own fate. The gods then came to resemble humans in their actions. In the Mesopotamian flood poem the Atrahasis Epic lower deities go on strike. They are exhausted from digging irrigation canals for the city-building higher gods. In response the mother goddess creates humans to take over the task.

The change was complete by the last millennium before the common era, and with its completion a great transformation began. As city dwellers came to rule as lords of the new empires the sacred began to recede. The experience of life's spiritual dimension no longer seemed as close or as accessible to the sophisticated inhabitants of great cities as it once had been to their ancestors. In the absence of the immediacy of the sacred, Armstrong claims, a kind of malaise infected these civilized city-states from China to Greece. With societies fragmented into class and trade, people lived far from direct contact with the sources of spirituality felt so closely in earlier epochs. New insights and a change in the mythic stance were once again required. Those changes led directly to the forms of religion and science we have today.

In China it was Confucianism and Taoism that embodied this change. Each contained a new emphasis on individual actions. It was through a human being's own effort, living harmoniously and in accord with a deeper underlying reality, that contact with the sacred was regained. In India Hinduism and Buddhism made a similar shift, placing the essential responsibility for spiritual life with the individual. Through meditation and right action each person could recover his or her place in an elemental unfolding of time. The monotheism of the Middle East also placed spiritual responsibility on the individual. Judgment came through each individual's acts. These acts would be known by the one true God who was both lawgiver and judge. In all these cases, from China to Israel, the individual's behavior and intention became the mediating factor determining access to the perceived sacred source of life.

In Greece a very different stance arose. Behind the physical world lay a realm of ideal forms. But these ideal forms were accessible. They could be known. They could be discovered through the effort of the mind. It was reason, the power of rational discourse, that allowed the structure of the physical world to be revealed. Mathematics and logic was the route to Truth. It was the contemplation of mathematics and the Truth it laid bare that revealed the world's highest domains. Through exploration of mathematics the Greeks felt they had discovered the world's essence and a new expression of the sacred.

Evolution, either physical or cultural, moves at different rates. Sometimes, in fact, it appears that evolution simply explodes. In the realm of physical evolution scientists have yet to understand why, five hundred million years ago, most of the phyla in the tree of life appeared in the space of a few hundred million years. This period, called the Cambrian Explosion, marks a singular event in the evolution of life on this planet.[13] The evolution in human culture and consciousness in the last millennium before the common era was a kind of Cambrian Explosion of the intellect and spirit. The philosopher Karl Jaspers has called this moment the Axial Age. Like the rapid development of body forms in the Cambrian Explosion, the Axial Age appears as a time of profound creativity. In the space of five hundred years the basis for all the world's "major" religions were set in place. With these religions came new hierophanies in the form of contemplative practices, rites, and symbols. The seeds of scientific practice were also sown, and, to their Greek creators, the mathematics found everywhere in the world was also taken as a new kind of hierophany.

The emerging religions did not do away with myth. The nature of sacred narratives simply changed once again. Each of the new religions had its own sacred stories, including the Buddha's awakening and the miracles of the Old Testament. Greek protoscientists saw stories in the form of a solar system centered on the Earth or of a universe that could be explained solely by the motion of atoms. Even in the Axial Age stories and descriptions of unseen and transcendent realities would be required. The need for the essential elements of myth was never lost.

The development of myths is an essential component of the evolution of human culture. The ubiquity of these myths and the apparent similarities of narratives found in widely separated cultures have been the focus of intense scholarship. As we have seen, the roots of science and religion are closely intertwined in the history of myth. Because of this proximity, the constellation of ideas emerging from the field of comparative mythology can provide fertile ground in our own explorations of science and spiritual endeavor.

Understanding myth has been likened to understanding an elephant by feeling its different parts. The tail seems like a snake, the ears like dry leaves, the legs like tree trunks. The study of myth over the past one hundred and fifty years has produced a variety of views on its nature, origin, function, and continued relevance. Joseph Campbell's insistence that myth reveals the collective, interior response to life is only one of these views.

THE PARADOX OF JOSEPH CAMPBELL

Indeed the first and most essential service of a mythology is
this one, of opening the mind and the heart to the utter
wonder of all being.
Joseph Campbell, *The Inner Reaches of Outer Space*

Campbell . . . attempts to generate a broad sympathetic
understanding of the role of [myths] in human history, in
our present day lives and in the possibilities for the future.
Kenneth Golden, *Uses of Comparative Mythology*

[Joseph Campbell] was neither a scholar nor a gentleman.
Wendy Doniger, *A Very Strange Enchanted Boy*

My head was full of Carl Sagan when I landed at the University of
Colorado as an undergraduate astronomy major. For people my age
Sagan was a hero and an inspiration. He was a scientist full of passion for the great work of deciphering the universe. To my surprise,
I quickly learned that my professors did not share this enthusiasm.
Invoking Sagan's name in an office visit with an instructor would often
be met with either "Oh God, not Sagan," or simply the raised eyebrow
that told me, "Go no further." Sagan, the premier astronomy popular-
izer, had detractors in his own field, a fact made painfully clear when
his application to join the prestigious National Academy of Sciences
was voted down. Sagan was a popular scientist who was often not popular with scientists. Joseph Campbell appears to share a similar fate in
the field of comparative mythology.

Campbell's writings inspired popular interest in mythology in the
United States. There can be no doubt of that fact. Campbell, a professor at Sarah Lawrence College for forty years, wrote prodigiously on
the interpretation and importance of myth. Many of his books remain
in press in popular editions decades after their publication. More than
his books, it was *The Power of Myth*, a PBS series featuring conversations between Campbell and Bill Moyers, that brought Campbell and
his ideas on the relevance of myth to millions. Many of those viewers

then became his readers. The principal idea that Campbell fashioned was the unifying themes of the world's mythologies. Myth, according to him, united us because it showed we had never really been separate.

The Hero with a Thousand Faces is Campbell's most famous book. In it he draws a term from James Joyce — *monomyth* — and uses it to describe universal patterns expressed in the heroic tales of every culture.[14] Campbell explored the hero myth from sources as diverse as the Arthurian legends of medieval England and the ancient Sumerian epic of Gilgamesh. In each story he found the same structure: a hero must undertake a dangerous journey for the good of his people and in doing so is himself spiritually transformed. Campbell felt he had found a fundamental unity in the world's religious underpinnings in the common elements of the monomyth. That unity helped articulate the ongoing relevance of mythic narratives for our own time. The hero's journey is everyone's journey. The myth of the hero's dangerous confrontation with the unknown was a description of the universal experience of growth and transformation that each person must undertake in his or her life.

Campbell drew heavily on the psychology of Carl Jung. In particular, he found the concept of archetypes, universal symbolic forms, to be the illuminating and effective source of myth. The world's mythic heritage gains its power by expressing itself through a collective storehouse of archetypes. These manifest in narrative themes, metaphors, and visual images. The potency of these myths and their archetypes are such that they did not disappear even as humanity entered the "modern" era. For Campbell the power of the monomyth was its enduring capacity to guide and instruct people through the many transformations of their lives. Universal myths as the guide for revealing universal meaning would be Campbell's theme throughout his life.

Campbell described this theme to Moyers:

> You have a body with the same organs and energies that a Cro-
> Magnon man had thirty thousand years ago. Living a human
> life in New York City or living a human life in the caves, you go
> through the same stages of childhood, coming into sexual maturity,

transformation of the dependency of childhood into the responsibility of manhood or womanhood, marriage, then failure of the body, gradual loss of its powers, and death. You have the same body, the same bodily experiences and so you respond to the same images. For example a constant image is that of the conflict of the eagle and the serpent. The serpent bound to the earth, the eagle in spiritual flight — isn't that something we all experience? And then when the two amalgamate we get a wonderful dragon, a serpent with wings. All over the Earth people recognize these images. Whether I am reading Polynesian or Iroquois or Egyptian, these images are the same and they are talking about the same problems.[15]

Campbell's vision is inspiring stuff, a grand description of both unity and pluralism. So why do many scholars of mythology raise their eyebrows at Campbell's name? To my surprise, I have found that Joseph Campbell is a popular scholar of mythology who is not popular with mythology scholars. Take, for example, the textbook used in my university's Introduction to Mythology class. It has a long chapter on different theories used in the interpretation of myth. Some of the great scholars in comparative mythologies appear: Edward Tylor, J. G. Frazier, and Claude Lévi-Strauss. Joseph Campbell, however, is not mentioned. As in the case of Carl Sagan, the issue may be jealousy or turf protection or simple narrow-mindedness. Unlike the case of Carl Sagan, the problem may also be that Campbell simply got too much of his story wrong. Many scholars of mythology believe Campbell's ideas do not match the evidence. In all forms of inquiry there is a difference between how one wants the world to be and what the data say.

Wendy Doniger, Mircea Eliade Distinguished Service Professor at the University of Chicago, criticizes Campbell for a shallow reading of myth in the service of his own ideas. "We must be grateful to Campbell," she writes, "for making so many people aware of the existence of the great myths . . . but we must regret that he did it so slickly that no one was ever encouraged to go on to the second stage. To do the hard work done by other[s], the work of understanding what the myths really say." Campbell's approach is, for Doniger, "a TV dinner of mythology. . . . [E]verything tastes the same."[16]

Some of the numerous attacks on Campbell seem overtly personal and shrill, making it difficult to separate sour grapes from spoiled logic. Doniger's insistence that Campbell could not see past his preconceptions, however, has the ring of truth. It is criticism that we must keep in mind as we make our way forward. Myth is the most ancient way humans have to create meaning and respond to experienced truths. To meet myth on its own terms and understand how it can still teach us about science and spiritual endeavor, we need both the long and the broad view.

Myth as Bad, Early Science

Theories of myth and its continuing relevance hinge on its collision with science. This is the view of Robert Segal, a comparative mythologist at the University of Toronto. Either myth has to be rejected because it has been supplanted by science, or it must be "regrouped." Regrouping means myth does not conflict with the ability of science to explain the natural world because myth does something very different from science.[17]

The first theories of myth were rejectionist. They saw myth as crude and naive attempts to explain the natural world. In the nineteenth century Max Muller, a philologist, claimed that myth was nothing more than a distorted description of the natural world. The personification of celestial phenomena into deities — what he called solar mythology — occurred through the "disease of language."[18] Mythic accounts of natural phenomena had originally been symbolic even to the ancients who created them. In time the symbolism was lost. People began to read myths literally because ancient languages had no abstract nouns or neutral genders. Thus by an accident of language the Sun became "he" and the Moon became "she" and both became deities.

The rejection of myth continued with the influential nineteenth-century anthropologist Edward Tylor. For Tylor myth and science were at odds because both were, fundamentally, descriptions of the physical world. The myths of "savages," as he called them, were a form of animism; that is, all aspects of the world were spirit laden. He called primitive religion "savage biology." There was nothing in myth other than crude attempts to explain the world. The replacement of the "animistic astronomy of the lower races" with the mechanical astronomy of modern science was simply a consequence of history and progress.[19] Tylor read myth literally and had no sympathy for those who would see myth as allegory or moralistic tale. The sacred narratives of indigenous people and of early cultures were accounts of how the world worked. They were to be rejected because the world simply did not work that way.

J. G. Frazer was a Scottish-born classicist whose monumental 1914 work, *The Golden Bough*, became enormously influential. Frazer focused on the myth of the Dying and Reviving God, which he connected to the common Paleolithic practice of ritual king sacrifice. Each year the king would be killed, usually symbolically, to ensure the next harvest's bounty. Science and myth also stand at odds in Frazer's view. For Frazer "myth is part of a primitive religion which is itself a counterpart of natural science."[20] While Tylor thought the purpose of myth was to explain events, Frazer saw it as a way to affect them. Through myth and ritual a mother goddess's underworld descent and eventual reappearance would be made real again. Myth was almost like a kind of technology. By enacting the myths through ritual, early cultures sought to ensure that the cycle of plenty would be renewed.

Myth and the Structure of Thought

For later theorists of myth the point was not bad science but the very nature of human thought. For some of these writers myth showed a fundamental difference between the "primitive" mind and our own. Myth could not simply be an early form of science because such thinking was impossible for "primitives." This was the view of the French scholar Lucien Levy-Bruhl. Mythic thinking for Levy-Bruhl was literally the opposite of scientific thinking.

The most important and influential scholar on myth as a mode of thinking was Claude Lévi-Strauss. In many of his works, including *The Raw and the Cooked,* Lévi-Strauss argued that myth reveals a kind of logic that is hardwired into our brains. An abstract structure occurs in all myths, and it, rather than the specific content of the story, is what really matters. For Lévi-Strauss there was no "inner core of meaning" in myth. Instead each myth shows us the structure of human conscious processes. Levi-Strauss's "structuralism" saw myth as a pattern of thought used to deal with the basic problems of human experience. Polar opposites are the all-important recurring pattern in myth: man and woman, light and dark, barren and fertile. For structuralists the symbolic language used in myths reveals our attempt to negotiate the complementary poles of experience. It is as if our minds were structured with binary logic. Myth's function is to show how these opposites can be mediated. Myth resolved the internal conflict. Thus for Lévi-Strauss myth is not about meaning; it is about mind.[21]

Myth, Experience, and the Sacred

Mircea Eliade wrote extensively on myth. His emphasis in these studies was always the primacy of human encounters with the sacred. Eliade regroups myth by giving it a function that is not directly in conflict with science. Myth does explain natural phenomena and social realities, but it does so by making the "wholly other," or sacred aspect of the world, apparent.

Eliade writes, "Myth narrates a sacred history; it relates an event that took place in primordial Time, the fabled time of the beginnings. . . . In short, myths describe the various and sometimes dramatic breakthrough of the sacred (or the 'supernatural') into the world. It is the sudden breakthrough of the sacred that really establishes the World

and makes it what it is. Furthermore it is as a result of the intervention of Supernatural Beings that man himself is what he is today, a
mortal, sexed and cultural being."[22] Myth for Eliade is always a sacred
narrative, and ultimately it is always about experience. Its intention is
not to simply explain but to renew experiential contact with the sacred
character of life.

"Renewal and regeneration" was the function of myth for Eliade.
Mythic narratives were only recounted at special times: rituals of ini-
tiations and the New Year. Those hearing the myth had to be in the
proper state to receive them. To hear a myth was be transported back
to the sacred time and space when the myth occurred. As Eliade writes,
"What is involved is . . . a return to the original time, the therapeutic
purpose of which was to begin life once again, a symbolic rebirth."
Myth generates experience and that is what made it a living force in
the life of early and indigenous cultures. It is an encounter with the
divine.[23] For Eliade, myth itself was a hierophany.

The power of myth was something humans could not live without.
According to Eliade, all the revolutions of the modern world have not
changed this fact. While we moderns think of ourselves as driven
primarily by rational and intellectual needs, Eliade sees myth in all
aspects of our culture. In books, plays, and movies we continue to
respond to its imperatives. As Eliade writes, "The cinema, that dream
factory[,] . . . employs countess mythical motifs — the fight between
heroes and monsters, initiatory combats and ordeals, paradigmatic
figures and images. . . . Even reading includes a mythological function because through reading modern man succeeds in obtaining an
escape from time comparable to the emergence from Time effected by
myths."[24] Thus for Eliade even though we believe ourselves above myth
we cannot sever our connection with the deeper forces that drive its
effectiveness. Our attempts to create such a disconnection is the very
source of the desacralization of the modern world that Eliade decries.

In his consideration of the continuing relevance of myth, Robert
Alan Segal writes, "Whether Myth has a future depends on its capac-

ity to meet the challenge posed by modern science."[25] Our short trip through comparative mythology shows different views on myth and science. For early writers like Tylor, myth was simply an early attempt at science. For structuralists like Lévi-Strauss myth can, perhaps, be thought of as data relevant to the science of human consciousness. For Campbell, Jung, and Eliade myth was not about science at all but something science did not address. Eliade, like both Jung and Campbell, regrouped myth, allowing it to coexist with science. Eliade, Jung, and Campbell felt that myth still functioned in modern human culture even if that function was denigrated and pushed underground. Jung and Campbell differed significantly from Eliade, however, in the focus on the unconscious. As Segal explains, for Jung myth is what reveals the existence of the unconscious. Through myth humans are brought into contact with the unconscious to foster psychological transformation. "One can never experience the unconscious directly but must experience it via myths and other symbolic manifestations."[26] Eliade viewed myth as allowing people to renew their contact not with the subconscious but with the sacred.

Much of the thinking about myth draws it into the domain of perceived spiritual realities. But science also appears as a background against which to understand its origins and functions. With these ideas in mind, and our own explorations of the hierophanies of science in the last chapters, we can now step forward to see how myth can be put into service to reenvision the relationship between science and spiritual endeavor.

THE STORY OF FREDRICK: OF NUTS
AND NARRATIVES, SEASONS AND STORIES

Once there was a family of field mice that lived within the cracks and crevices of an old garden wall at the edge of an abandoned farm. During the first year after the farmer left the mice were anxious at the turning of the seasons. As summer turned to fall the mice knew that without the farmer and his grain a cold, dark, and hungry winter stretched before them, so they set about gathering what they could: nuts and corn, straw and berries. They scurried about the fields, as mice do, grabbing what they could, working night and day to prepare for the winter. One member of the family, though, remained apart from the group. Fredrick, the smallest of the mice, spent his days on the garden wall staring, it seemed, at nothing.

One day the other mice came to Fredrick. "Why aren't you working, Fredrick?" they asked. "But I do work," said Fredrick. "I gather sun rays for the cold, dark winter days." The other mice turned away. They were cross, and they shook their heads in anger.

Later in the fall they saw Fredrick stitting quietly on a high rock watching over a field of yellow grass. "What work are you doing now, Fredrick?" they demanded. "I am gathering colors," Fredrick said without anger or disdain. "Winter is grey and we will need them."

Later that day Fredrick seemed to be asleep on his perch. "Now what are you doing while we work, Fredrick?" they said. "Are you napping?" Fredrick was not asleep. His eyes were bright and he responded with kindness, "No my beloved ones, now I gather words. The winter will be long and we will run out of things to say."

Soon the winter came with its strong winds and driving snow. Inside the wall the mice dined happily on the great store of food they gathered. They felt happy and secure. The days turned to weeks and into months. By February most of the corn was gone and the sweet taste of the berries seemed like a bygone dream. The mice were cold and hungry, and they fell into despair.

Fredrick saw their sadness and stepped forward. "Remember," he said, "I gathered my own supplies. I will share them with you now." He cleared his throat and spoke with a strength that surprised his family.

"I send you the rays of the sun. Do you feel their golden glow?" In their mind's eye the mice could see the sun and suddenly their fur felt warm, as if they were standing in a sunbeam on a beautiful spring day. Then Fredrick brought them his colors, painting a vision of rapturous spring flowers: periwinkle, daffodils, and lilacs. The little mice delighted as the colors appeared before them and exploded in their hearts like fireworks. "You said you gathered words too," one of the mice said. "What of the words, Fredrick?" Fredrick stood tall on his perch and began.

"Who scatters snowflakes? Who melts the ice?
Who spoils the weather? Who makes it nice?
Who grows the four-leaved clovers in June?
Who dims the daylight? Who lights the moon?

Four little field mice who live in the sky.
Four little field mice . . . like you and I.

One is the Springmouse who turns on the showers.
Then comes the Summer who paints all the flowers.
The Fallmouse is next with walnuts and wheat.
And Winter is last . . . with little cold feet.

Aren't we lucky the seasons are four?
Think of a year with one less . . . or one more!"

When Fredrick was done, all the mice embraced him. "Thank you" they said, and Fredrick smiled.[27]

. . .

I have paraphrased Leo Lionni's delightful little story of Fredrick the poet fieldmouse. His own version is better than mine and worth reading to your kid or just for yourself. It's a story that is used a lot in grade school classrooms. Teachers read the story to their charges and then ask them what Fredrick had contributed to his little family. The kids get it, of course. As one second-grader, Tiffany C. of New York, wrote in her online review of the story, "I think you should read this book because you should find out what Frederick was working for. I think you should read this book because Frederick is really working hard, so you should read it to see what it's about."[28]

Lionni's story teaches children the importance of art and poetry. It teaches them that there is nutrition in words and stories. I once overheard a parent mutter the following cynical response to Fredrick's story: "Yeah, but given a choice I'll take the grain. Maybe somebody should have pushed Fredrick out of the hole in the wall and let him freeze." This response struck me as being a lot like that of scientists who cannot see the equivalent roles of Mythos and Logos, who see spiritual endeavor as something for people with inherently weak minds who cannot face the hard facts of the cold, purposeless universe revealed by research.

I think they are deeply, woefully wrong. The fires of spiritual endeavor burn from the same fuel as those that animate scientific pursuit; and both were ignited first in the universe of myth. Missing this, they miss the roots of their aspiration and can see only half of its importance, its power, and its possibility.

SCIENCE, MYTH, AND THE SACRED

From our ancestors rising upright three million years ago to the birth of industry in coal-stoked fires, we have changed in biology and culture. From the first narratives of ascent to stories of mother goddesses descending to Hades to the new architect gods of our first cities, our myths have transformed with us. Observation and deeply felt experience were channeled into these stories, providing meaning, guidance, and instruction. But in the twentieth century it has been our *understanding* of myth that has changed. From Robert Tylor's early dismissal of myth as primitive science to Mircea Eliade's hierophanies to Joseph Campbell's *Hero with a Thousand Faces*, we can see the possible routes for myth's continuing relevance. Myth, it seems, remained with us even as we rejected it. We cannot do without these narratives because we cannot do without stories that connect us to our deepest sources of inspiration. The constancy of our imperative to create meaning through myth illuminates it as a kind of connective tissue drawing together science and spiritual aspiration.

Myth is the taproot from which science and religion grew. The origin of the world's major religious lineages coincided with the creation of science's formal roots in Hellenistic Greece. This co-incidence or co-appearance is striking. It marked the beginning of a split that would take another two and a half millennia to come to fruition. For at least forty thousand years human cultural evolution was saturated with myth. Those myths expressed our first organized response to the religious experience described by William James and Rudolph Otto. Myth was our first response to the world's character of awe, wonder, and power. It was also a record of our first organized investigations, as shown in the astronomical orientations of locations such as Newgrange. In myth and ritual we find our first response to the sense of the world as sacred, and in that response the impulse that would become science and religion were carried.

Whatever myth and rites were enacted at Neolithic monuments in Newgrange six thousand years ago, they clearly involved close observation of astronomical patterns. Whatever deeper meanings the myth of the king's sacrifice had for agricultural cycles, it involved close observation of those cycles. Myth *was* a response to the natural world. It was a response to our investigation, our questioning of that world and our place within it. In that understanding there was also the attempt to control the world for our own ends. This is the seed of truth that early theories of myth captured. Myth was not simply primitive science, as Tylor argued, but it did embrace the aspiration and activity we now identify as science.

There are different kinds of myth. Some speak only to the internal life of humans as they make their transitions to adulthood, parenthood, and old age. Others speak of the external world, the universe of origins and endings. They tell of why the Moon has phases, where the yearly cycle of plant life originates, and how the mountains came to be. While early theories of myth as primitive science could not tell the full story, purely psychological accounts of myth miss the point that some myths arose from rapt attention to the movements of the natural world. Not all mythic stories relate to the natural world, but those that do make the connection to the world's sacred character. As Eliade explained, myths were hierophanies. In their telling, the listener had to be receptive to the story's power and its ability to reanimate the origin times. An audience gathered for the ritual recounting of the great stories embraced by myth would encounter space and time as open and reborn. The narrative was made real and fully present.

Is there anything different happening in a popular astronomy lecture on the Big Bang? In the widely popular *Cosmos* series, why was Carl Sagan shown endlessly standing before his imaginary spaceship's view screen staring in wonder as the swirling galaxies glided by? The narratives of science have the capacity to transport us into states of awe, wonder, and even rapture. Again and again in popular science books, *NOVA* specials, and science-fiction films, the stories of science are called forward to invoke a sense of the world's grandeur and power. These are the hallmarks of sacred narratives, and the only reason we do not call them such is, I think, a kind of collective amnesia.

Scientific narratives are mythic narratives because they have their roots in an aspiration and activity we have carried forward since our own origins. Let us be clear, however: the maturation of the scientific process over the past two millennia and especially the past five hundred years has refined it into a uniquely powerful tool for investigation. That maturation has led to deeper understanding about "proper" methods of inquiry that our ancestors had not worked out. At the same time the importance of storytelling in science cannot be overstated, and this is an essential part of its connection with myth. All the grand challenges in science form narratives: the origin of the cosmos, the origin of life, and the evolution of humans. Even questions about specific natural processes, gene expression, the emission of light from atoms, must still be communicated as narratives. First "this" happens, and then "that" happens. First this element does one thing, and then a second element does another. In science we are always telling stories. Even a scientific paper with the boring title "A New Riemann Solver for Magnetohydrodynamics" (from my own field) is a story with its trials and errors, discovery and success, and, finally, usefulness in the broad cooperative effort.

Every culture has its system of myths that provides meaning and context. As Armstrong and others have written, myths only work for a culture when they continue to be useful, when they remain successful in creating meaning. If one could ask a Neolithic farmer what the world was made of, his response would likely be couched in the framework of his myths. Likewise, ask a person on any city street in the United States the same question, and you would hear about atoms, electrons, stars, planets, and galaxies. Regardless of scientific literacy, these words form the background of our understanding.

In telling its stories, science forms a mythic system for our culture. The success of science and its fruits in technology provide obvious evidence of its power. But beyond jet travel and antibiotics lies the power of science's narratives to continue opening the gates to the original experience of the sacred. We do not think of science as playing this explicit role. We do not emphasize this capacity or teach it to our students. Nonetheless, it is there. This capacity for scientific narratives to act as hierophanies, to serve as gateways to an experience of the sacred, has been muted and distorted. As we have seen in the first chapter, this occurred through the specifics of history over the past few centuries. In that time science was drawn into a polarity with spiritual endeavor, which, perhaps, it never needed to take. More important, even if that polarity between science and spiritual endeavor was necessary to allow science to mature, the time has come to see past history to prehistory. The time has come to reimagine science and religion in terms of their common mythic roots and continued mythic necessity.

Myths will be needed as long as human beings exist to create them. The tragedy of our era is that in attempting to become fully rational, we did not kill myth; we simply deformed it and ourselves in the process.

There is an imperative to recovering the mythic response within the activity and understanding of science. The desacralized world that Eliade spoke of is our world. It is full of marvelous inventions and life-prolonging technologies. It is also dead to the moorings of interior meaning that humanity requires to make sense of its deepest experiences. Myth has always been the vehicle that makes these meanings manifest. That was Jung's insight. Through myth our place in the universe, our experience of change and death, is placed in the context of a sacred, indefinable essence of Being. That was Eliade's insight. These moorings have never been cut until now.

After uncounted eons of human beings living within and through their sacred narratives, in the past few centuries we have conducted an experiment in consciousness wherein the mythic understanding of life and the cosmos has been rejected. The emphasis on the rational empirical approach to life has certain benefits, but this is only one aspect of human being. A century of barbarism on mass scales and environmental decay so profound that the climate itself appears to be altered stands as stark evidence that our modern encounter with the world cannot be counted as uniformly "better" than those we see in the rearview mirror. Humanity still needs myth to interweave the exterior world and our interior response.

Some people look at the dangers we face and the flattening possibilities of experience and say, "Science is the problem." This is a foolish response. There is no going back to some prescientific Eden short of a catastrophe that no one wants personally to be part of and that is unlikely to be much of an Eden. Indeed, the material benefits of the scientific worldview are too numerous to simply reject. But what is required now is wisdom, and that can come through a vision of science not divorced from the universe of myth. Science grew from that universe and retains its capacity to connect us with the living roots of an experience of the sacred. In the next section we will explore two examples of great themes in science, cosmology and climatology. We will tell each story and unpack each one's mythic roots. This will give us concrete examples of the continuing relevance of myth as expressed in science and science's capacity to serve as gateway to the interior universe of the sacred.

PART II

The Terrain

Sacred Narratives in Science and Myth

CHAPTER 6

The Origin of Everything

*Big Bangs, the Multiverse,
and the Parade of Ants*

If you're religious, it's like seeing God.
George Smoot,
Principal Investigator for the Cosmic
Background Explorer (COBE) satellite

The real business of poetry is cosmology.
Robin Blaser, *The Fire*

Courage is lost in the wild dark hours when chaos
swirls, and face to face with the abyss, you near the
white fire of time.
Ellen Hinsey, *The White Fire of Time*

Indra was king of the gods. Brave, noble, possessed of a compassionate heart, he looked after the domains of both the divine and human worlds with the steady hand of a wise father. When a great dragon broke free of its prison and left the realm of men and gods in devastation, Indra stood alone as the world's last, best hope. The dragon had swallowed the waters of heaven and lay immovable at the top of a mountain. Indra

141

called to the dragon to release the waters, but the beast refused. The battle that followed was terrible, shaking the cosmos to its very roots. When the dust settled Indra's victory over the beast was as complete as it was glorious. The waters were freed. They streamed in ribbons across the land, flowing once again through the world.

"This flood is the flood of life and belongs to all. It is the sap of field and forest, the blood coursing in the vein. The monster had taken it for himself, away from the world. But now it is released."

During the reign of the dragon the city of the gods had crumbled and cracked. Indra's first act was to rebuild the mansions and halls of the great city. He summoned Vishvakarnan, master of the arts, to build a palace befitting the king of the gods. Vishvakarnan created a shining residence, marvelous with gardens, lakes, and towers. Indra was not satisfied. He wanted more of everything. "Give me bigger ponds, trees, towers, and golden palaces!" he demanded. Whenever Vishvakarnan was done with one thing Indra wanted another. The divine craftsman fell into a deep despair. He complained to the Brahma, the Universal Spirit, who abides far above the gods. Brahma comforted him: "Go home; you will soon be relieved of your burden." Brahma then approached Vishnu, the Supreme Being of whom he, Brahma, was but an agent. Vishnu listened and nodded his head.

Early the next morning a Brahmin boy appeared at the gate of the palace asking to see the great Indra, king of the gods. The boy was slender, ten years old, and radiant with wisdom. The king welcomed the boy with gifts of honey, milk, and fruits. "Tell me, oh venerable boy, why did you come here?" asked Indra. The beautiful child replied with a voice that was as deep and soft as the slow thundering of rain clouds. "Oh, King of the Gods, I have heard of this palace you are building and have come to ask you some questions. How many years will it take to finish this rich and extensive residence? Surely no Indra before you has ever succeeded in completing such a task."

Indra was amused by the boy. How could this child have known any Indras other than himself? "Tell me, child!" he said in a fatherly manner. "How many other Indra's have you have seen — or heard of?"

The boy replied in a voice as warm and sweet as milk from a cow but with words that sent a chill through Indra's veins. "My dear child," said the boy. "I knew your father, the Old Tortoise Man, progenitor of all the creatures on the earth. And I knew your grandfather, Beam of Celestial Light, who was the son of Brahma. Also I know Brahma, brought forth by Vishnu from a lotus growing from Vishnu's navel. And Vishnu, too, the Supreme Being, I know."

"Oh, King of the Gods, I have seen the dreadful dissolution of the universe. I have seen it all perish again and again, at the end of each cycle. At that time every single atom dissolves into the primal pure waters of eternity, whence originally all arose. Who will count the universes that have passed away, or the creations that have risen afresh, again and again from the formless abyss of the vast waters? Who will search through the wide infinities of space to count the universes existing side by side, each containing its own Brahma, its Vishnu, and its Shiva? Who will count the Indras in them all?"

As they were talking a procession of ants had made its appearance in the hall. In military precision, the tribe of ants paraded across the floor. The boy noticed them and suddenly laughed. "Why do you laugh?" stammered Indra. The boy answered: "I laughed because of the ants." The king pleaded, "Why? What is it about the ants that makes you laugh?" The boy drew himself up, and with his eyes fixed on Indra he said, "I saw the ants file by in a long parade. Each of them was once an Indra. Like you, each by virtue of his deeds ascended to the rank of king of the gods. But now through many rebirths each has become again an ant. This army of ants is an army of former Indras."

The king of the gods was speechless. The boy turned and left. After many days alone Indra called his architect and thanked him for his work. "You have done enough," Indra said. "You may rest now."[1]

· · ·

In the beginning there were always beginnings. Every culture has cosmological myths. Every culture has narratives that explain and describe what the universe is and how it came to be. Each of these stories manifests a grave hierophany. Each speaks of deep mysteries, the very essence of time and space. To hear these myths in their proper time and place was to serve as the world's memory, recalling the timeless acts from which time was born. When these stories were told the listeners became participants in the Creation, the most sacred of all endeavors cosmic or human.

Modern cosmology is our culture's origin story. It is the science of the universe entire, and its purview embraces the origin of all space, time, matter, and energy. With so large and so wide a charter it is no wonder that cosmology harbors the most compelling links with the previous universes of myth. That makes it a good place for us to begin. In the previous chapters we have looked at science, religion, and myth in isolation. Now we can explore examples of how science fulfills the role of myth while retaining living roots in mythological imperatives. By providing narratives that renew our connection to the innermost experience of awe, science continues an ageless tradition that stretches far into prehistory.

Cosmology is a new science. After centuries of speculation our ability to meaningfully address cosmological questions represents a recent development in our scientific capabilities. The real and profound success of cosmology is a tribute to the capacity inherent in modern science. Scientists must simultaneously address the very large and the very small in their cosmological studies. The world's most powerful scientific machines must be yoked to the endeavor: telescopes situated on mountains so high that astronomers must risk altitude sickness to carry out their research; giant particle accelerators so powerful that they bring us back to bare instants after the moment of creation. Through the perspective of cosmology we have come to see a universe that is no mere empty box filled with stars and galaxies. Instead the whole of creation becomes a fabric woven of space and time. Our familiar three-dimensional world becomes just a thin slice of a much richer world of possibilities, linked to the creation of many, perhaps infinitely many, more universes. As I write this book the narratives of cosmology are changing. Where once we thought only of the Big Bang and a single universe, now we must consider the stranger world of the multiverse.

We will now explore the narratives of modern cosmology beginning with Einstein's theory of relativity and classic Big Bang theory. From the Big Bang we will follow the story forward as the science responds to new discoveries and rising paradoxes. The theory of cosmological inflation embodies one such response. This is now the dominant modern paradigm for the origin of the universe. Ironically, from inflation emerges not just one universe but many. The multiverse — a universe of universes — appears as a consequence of the new cosmology. Such a cosmos of endless universes might have seemed as familiar to the Vedic authors of Indra's myth as the classic Big Bang does to the Judeo-Christian West.

Always in telling these scientific stories we must keep our eyes on their mythic dimension. For they are the fruit of the creative process in science as researchers respond to the necessities of data and theory. But imagination plays a vital role in that response as well. In these cosmological stories there is an echo of something older that we can strive to hear. The narratives of cosmology, like those of many other sciences, can connect us with our submerged roots of experience. These are the experiences that William James would have called religious and that were the explicit purpose of myth.

EINSTEIN AND THE SHAPE OF SPACE-TIME

> Space by itself, and time by itself, are doomed to fade away into mere shadows, and only a kind of union of the two will preserve an independent reality.
>
> Hermann Minkowski

Cosmology was waiting for Einstein. He made it possible. There had been early attempts to build scientific models of the entire universe.

Newton had certainly thought about it, as had Immanuel Kant.[2] These efforts, however, were hobbled without Einstein's great insight into the nature of space and time as a whole.

The problem is deceptively simple. Scientific cosmology requires a theory of the universe as a whole. Here "theory" means a complete mathematical description. A mathematical model of the universe must fully describe everything in it at every location and every moment in history. The model must also make predictions. It must tell scientists what to expect when they make their observations. Cosmology demands a testable account of the universe as an entity in itself. The point of cosmological science is to treat the universe like anything else in physics: an atom, a rock, a cow. But the universe contains all atoms, all rocks, all cows, and all astronomers. It not only contains everything like a giant box; it *is* the box. Cosmology begins with the premise that there is only one universe and we are inside it. But how can you describe everything from the inside? Einstein found a way.

The first step came in 1905 when Einstein published his special theory of relativity. In this remarkably short paper he did away with Isaac Newton's three-hundred-year-old vision of time and space. Space, for Newton, was just an empty stage on which the drama of physics was played. He considered it "absolute" in the sense that space constituted an identical emptiness everywhere. It was always the same, at all times. Time was also absolute for Newton. In his physics time was like a cosmic river flowing at the same rate everywhere in the universe without variation. No matter where you and I might be relative to each other, one meter for me was always as long as one meter for you. One minute on my watch also had to be one minute on yours.

Einstein swept away the river and the stage. In a single stroke he showed that space and time were neither separate nor separately absolute. Instead, each was malleable.[3] Space and time could each, separately, shrink or expand depending on the relative motion of the observers who measured them. One minute on my watch might last ten thousand years for you if I traveled on a spaceship at close to the speed of light and you stayed on Earth. One meter measured with my ruler as I fly by might be the width of a hair when you measure the same object. Space and time were not absolute. Each could change depending on the state of motion of the observer. Most important, physics could account for the change. The equations Einstein derived showed scientists how to relate the space and time different observers experienced. The malleability of space and time was a radical idea that did away with thousands of years of "common sense."

Einstein replaced space and time with a new entity that he called space-time. The three dimensions of space were merged with the dimension of time to create a new four-dimensional substratum of physics. Einstein's vision enlarged reality beyond the three-dimensional world of chairs, tables, rooms, and buildings. This new universe was a *hyperspace*, a world with an extra dimension compared with the usual up-down, forward-backward, left-right directions we are so familiar with. The consequences of this change were profound. In relativity every object becomes four-dimensional as it extends through time. You and I each trace out paths in this hyperspace beginning with the moment of our birth and stretching out until we die. What we see of each other is simply a kind of 3-D shadow of our extended 4-D selves. Einstein's relativity mapped out a way of understanding how those shadows would be cast. The new 4-D space-time was a fundamental shift in the way physicists understood reality. It was also the new stage on which the game of cosmology could be played. But first gravity, the most powerful of all cosmic players, would have to be reimagined.

It took Einstein another seven years of intense effort to complete his general theory of relativity and replace Newton as the lord of gravity. In Newton's physics, gravity was a force that mysteriously reached out between massive objects and pulled them toward each other. Einstein recognized that gravity could be understood not as a force but as a distortion of 4-D space-time. Space-time constitutes a kind of stretchable fabric, like cosmic rayon, that underpins our perception of events. It can be bent and distorted, as well as stretched. How much the fabric of space-time stretches or bends depends on how much matter is present and how it is arranged. The more matter present at a point, the more space-time stretches "into" that point and the more other objects with mass have to respond to the stretching. What had been an empty stage of space and time became a principal actor in the gravitational theater of physics.

All this talk of bending and warping is really about the shape of space-time. That in itself is a remarkable and radical idea. Once the distribution of matter was known, the equations of general relativity could describe the shape, that is, the geometry, of space-time. The equations describe all of it, everywhere. That is their beauty and their power. Einstein's years of work had not only given scientists a new description of gravity but also the tools to describe the universe as a whole. He had given them the equations of cosmology.[4]

It did not take long for Einstein, and those who followed him, to begin constructing cosmological models. These were mathematical descriptions of possible space-times, that is, possible universes. No data existed yet to test any predictions. It was a kind of theoretical game at first. Einstein's first attempt rested on his bias that the universe had to be static and unchanging. This was an old prejudice. It was hard for many scientists and philosophers to imagine a universe that evolved. Evolution implied a beginning and, perhaps, an end. Even without biblical inclinations many researchers were uncomfortable with any form of cosmic beginning. An eternal, unchanging universe seemed simpler and more elegant.

Using the equations of general relativity, Einstein first constructed a universe in which the gravitational "pull" of matter was balanced by an antigravity "push" that balanced the cosmos and made it static. Einstein got his repulsive antigravity via an entity he added, ad hoc, to his equations called the *cosmological constant*. Without the cosmological constant, Einstein's model universe would have collapsed under its own weight. Other scientists were not bound by the same preconceptions and had no need for a cosmological constant. In 1916 the Dutch physicist Willem de Sitter began with a different set of assumptions. Also using the equations of general relativity, de Sitter constructed a universe that was neither static nor unchanging. It expanded like stretched taffy. Every point in de Sitter's space-time receded from every other point as time advanced. Other scientists soon joined the fray, creating different theoretical universes with different properties. After about fifteen years of this abstract theorizing and mathematical gamesmanship the universe finally spoke for itself.

In 1928 the astronomer Edwin Hubble, using the most powerful telescope of his day, found that every galaxy in the sky was moving away from us.[5] The more distant a galaxy was from our own, the faster it appeared to be rushing outward. This is exactly what observers riding on debris from an explosion would see. In an explosive release of matter all the bits of shrapnel appear to move away from all the others. If you are riding on one bit of debris, the pieces that are farthest away from you are the ones that appear to have covered the most ground since the time of the explosion; thus they seem to have the highest velocity. In this way the interpretation of Hubble's data was straightforward. The universe was expanding and the galaxies were going along for the ride. Einstein, it seemed, had been wrong. The universe was not static. The science of cosmology had officially begun in earnest.

LET THERE BE THE BIG BANG

> Tune your television to any channel it doesn't receive, and about 1 percent of the dancing static you see is accounted for by . . . the Big Bang. The next time you complain that there is nothing on, remember that you can always watch the birth of the universe.
> Bill Bryson, *A Short History of Nearly Everything*

> The magic rat drove his sleek machine over the Jersey state line.
> Bruce Springsteen, "Jungle Land"

New Jersey seems an unlikely place for the origin of the universe to reveal itself, but that is exactly where the story of the Big Bang starts.[6] In spring 1964 two Bell Lab scientists, Arno Penzias and Robert Wilson, managed to trip over the greatest cosmological discovery ever made and win a Nobel Prize in the process. At the time the two astronomers were working on the new technology of microwave communications. Together they had constructed a massive horn-shaped antenna in a field outside of Holmdel, New Jersey. Their interest was transmissions to and from orbiting satellites that were just beginning to populate the sky. Unfortunately, the project wasn't going well.

The problem was an annoying, low level of "noise" that persisted regardless of which direction the antenna pointed. It was a microwave hiss that refused to go away. For weeks Wilson and Penzias struggled to root out the problem. They rebuilt the electronics. They cleaned layers of pigeon guano from the antenna surface. Nothing changed. Then, slowly, through painstaking work they came to understand the problem. There was no problem. The signal was real, and it was really, really old.

The microwave signal Wilson and Penzias captured wasn't noise but the ultimate prehistoric relic. The antenna they had built to communicate with Earth-orbiting satellites was, instead, picking up fossil signals left over from our universe's childhood 13 billion years ago. The light waves flooding their antenna were created a mere 300,000 years after the moment of creation.[7] This "cosmic microwave background" (CMB) filled every sector of the sky, in every direction. It was a pervasive electromagnetic memory of the universe's origin and a direct link to the time of the Big Bang. The serendipitous discovery of CMB radiation became a turning point for cosmology.

Before Wilson and Penzias's discovery astronomers had fallen in with the so-called steady-state model.[8] They had digested Hubble's result that all the galaxies were rushing away from each other, and you might imagine that would naturally lead them to conclude the universe had a beginning, a time when the expansion began. That conclusion, however, was still too radical for many astronomers. Many scientists still wanted a cosmos that *had* existed forever and *would continue* to exist forever. While the universe was clearly not static, perhaps, they argued, it was in a "steady state." To maintain the cherished notion of an unchanging cosmos they invented a "continuous creation" model in which new matter was slowly added to the universe, allowing it to expand forever and forever look the same. A few atoms of hydrogen appearing here and there were enough to make new galaxies in the expanding space and have it all work out. The champion of this steady-state model was the British astronomer Fred Hoyle, an irascible but brilliant scientist. Throughout the 1950s the steady-state model was the dominant vision of cosmology. Hoyle was so sure of its truth that he dismissed theories with a beginning, a $t = 0$, deridingly referring to them as "Big Bang" models.

With the discovery of CMB radiation, Hoyle's "diss" became *the* center of discussion. Suddenly the Big Bang made sense because the whole sky was glowing in microwave relics of the primeval explosion. Astronomers knew how to interrogate these light waves and to discern conditions in the matter that created them. The pattern of microwave energies in the CMB told astronomers that the light emanated from a very hot, very dense gas of atoms that must have filled all space. That inescapable conclusion was enough to kill the idea of the steady-state universe. Look out at night, and you can see for yourself that space looks pretty empty. It's just some stars and some gas clouds and then long stretches of nothing. That is the way things are now. The micro-wave background was telling astronomers that at some point in the past the universe was very different.

Imagine the expansion of the cosmos running backward like a movie in reverse. The space between galaxies shrinks, and soon galaxies are packed ever more tightly together. Run the movie far enough back, and all the galaxies and their stars eventually merge, dissolving into a dense soup of protons and electrons, the basic building blocks of atoms. This gas would be extremely hot, a natural consequence of jamming so much stuff together in such a small space.

A universe of hot, dense gas is exactly the requirement for making CMB photons. *Exactly* is the operative word here. The mathematics that describes radiation emerging from hot, dense stuff is very explicit in its prediction. An exquisitely simple description of the pattern of detectable microwaves falls right out of the physics. The microwaves Wilson and Penzias had stumbled on matched this prediction with stunning accuracy. The fit between the data and prediction is now so good that it is the equivalent of firing a rifle and hitting a target a kilometer away. That is good accuracy when you talking about a theory of the universe as a whole.

The radiation that Wilson and Penzias discovered proved that the universe had evolved. Most important, it proved that the universe had a beginning. Now the task before scientists was to push the clock back ever further and understand what that beginning meant. How had the universe started?

THE BIG BANG TRIUMPHANT

Seek simplicity and then distrust it.
Alfred North Whitehead

By the time the decade of disco and punk rock was well under way, the Big Bang had become the standard, universally accepted model for cosmic history. Scientists could use Einstein's general theory of relativity to tell them how the entirety of space and time evolved. That was the big picture. Into this dynamic cosmos they poured their understanding of particle physics, the science of matter at its most fundamental level. That was the small picture. Together general relativity and particle physics spun a grand narrative of cosmic evolution. The story goes something like this.

In the beginning there was a single geometrical point containing all space, time, matter, and energy. This point did not sit in space. It was space. There was no inside and no outside. Then "it" happened. The point "exploded" and the universe began to expand.

That opening sounds pretty good, but in point of fact the earliest part of this history is still unknown. Astronomers and physicists cannot start the story exactly at the beginning. Our physics is not good enough to go all the way back. Instead we can only start an instant afterward. There is a gap between the beginning and the beginning of the story. That gap has continually gotten shorter as scientists understand space, time, and matter more fully, but still the gap exists.

By the 1970s physicists certainly understood enough to set the clock running a few hundredths of a second after the moment of creation.[9] The universe had a temperature of 100 billion degrees at this point and was so dense that a single teaspoon of cosmic matter would weigh more than a thousand tons. As strange as it sounds, this was a universe physicists already knew how to work with. General relativity gave a complete and accurate description of cosmic expansion, and the equations of nuclear physics had already been vetted in particle accelerators for more than sixty years. Running forward from this point, the universe's life could be described with astonishing detail.

At one one-hundredth of a second after the Big Bang, the entire universe was about the size of our solar system. It was a universe pervaded by dense, primordial gas: an ultra-smooth, ultra-hot sea of protons, neutrons, and other subatomic particles. No atomic nuclei existed yet.[10] If the nucleus of an element such as helium or carbon did manage to form via thermonuclear fusion, it would instantly be smashed apart by other collisions. In this dense soup photons, which are quantum particles of light, mixed easily with matter. The photons were constantly emitted and absorbed by the roiling mix of matter particles. Each photon lived only for an instant and never traveled far before being gobbled up by a proton, neutron, or electron.

As the cosmic clock ticked off the instants expansion continued to stretch space, and with it the particle-photon sea thinned and cooled. For a brief period the primordial gas passed through the right temperature for sustained nuclear reactions. Protons and neutrons collided and combined to form nuclei of light elements such as helium and lithium. This was the era of "Big Bang nucleosynthesis," when most of the universe's helium was forged (think about that next time you buy a balloon at the circus). This cosmic nuclear furnace stopped just three minutes after the Big Bang when the universe dropped below the temperatures and densities at which nuclear reactions can be sustained. At this point all creation was a mix of photons, protons, electrons, and light nuclei. Photons of light were still tightly coupled to matter. They still could not travel far without being absorbed by a matter particle. But the time was coming when this intimacy must end.

After 300,000 years of expansion and cooling, negatively charged electrons were moving slowly enough to get caught by positively charged protons. Each capture created a new atom of hydrogen. Once the process started, the universe rapidly made the transition from a mix of free protons and electrons to a vast gas of electrically neutral atomic hydrogen. This was a critical moment in cosmic history.

Photons did not easily mix with the neutral hydrogen atoms. The cosmic bath of photons, so closely coupled with matter up to this moment, was suddenly frozen out. Because the photons could not easily be absorbed by hydrogen atoms, the matter-photon dance stopped. The photons were left orphaned. The universe was suddenly transparent to their travels. They could now cross the whole of space unimpeded. The light (i.e., photons) emitted by matter at this time of "decoupling" was doomed to wander the cosmos forever. As space expanded their wavelengths stretched until, eventually, they became the microwaves we now see as cosmic background radiation. The CMB, so critical for our knowledge of cosmology, was born.

The creation of CMB photons is important for many reasons. Up to the moment of decoupling, the cosmic sea of particles had been smooth, like a hyper-pureed soup. Any large lumps or clumps of matter were instantly smoothed out by the rapid motion of the particles and the pressure of the lightwaves. Nothing can be perfectly uniform, of course. Tiny waves in the cosmic gas rippled back and forth, producing regions minutely denser or more rarefied than their surroundings. After the cosmic background radiation decoupled from the matter, light could no longer act to smooth out the lumps. Gravity began to grab at these tiny ripples. Although the universe as a whole expanded, these overdense regions began to contract via the power of their own gravity. Soon they drew surrounding material into their orbit. The clumps grew ever larger. The growing blobs of gas are the seeds that will, in time, become the first stars and galaxies. The moment when the CMB photons decoupled from matter marks the beginning of an epoch of cosmic structure formation.

As the eons passed a vast cosmic network of form emerged from these humble beginnings. One by one a hierarchy of cosmic shapes was born. Galaxies appeared first. Then clusters of galaxies were swept together by their mutual gravitational pull. Finally, clusters of clusters of galaxies condensed out of the background sea of gas. In the 13 billion years since the Big Bang, gravity alone constructed a cosmic architecture that is filamentary and beautiful. Appearing as a foam of galaxies stretching across billions of light-years, large-scale structure, as astronomers call it, gives testimony to the enduring workings of basic physical law across the vast reaches of cosmological space and time.[11]

In their study of the large-scale distribution of matter astronomers came to an astonishing conclusion. The cosmic network or form is not just made of normal matter — the protons, electrons, and neutrons you and I are made up of. Instead, the universe is also composed of tremendous quantities of something else, something that emits no light. Detectable through its gravitational influence, this "dark matter" constitutes the majority of mass in the universe. While "normal" matter accounts for just 30 percent of the cosmic mass, it is dark matter in its preponderance that sculpts the large-scale structure we observe with our telescopes. Scientists still do not know what this dark matter is made of, but the evidence is all but overwhelming that it exists. The visible galaxies we see strewn across space are nothing more than strings of luminous flotsam drifting on an invisible sea of dark matter.[12]

So runs the short-form narrative of the Big Bang. Embedded in an ocean of dark matter, generations of stars form within the luminous galaxies. Planets form in swirling disks of gas and dust surrounding the stars, and, on at least one of these worlds, life begins and evolves. It grows intelligent and grows curious. It watches and responds with stories. After 13 billion years the universe generates its own eyes.

A BIG BANG FULL OF HOLES

Everything popular is wrong.
Oscar Wilde

The Big Bang model circa 1980 was a triumph of science. Researchers could finally say something about cosmic history that could be tested. But, as the observations became more refined and the theory was explored in ever greater detail, troubling dilemmas arose that threatened to topple the edifice of Big Bang cosmology.

The pressing questions facing the Big Bang model came from a variety of directions. Some questions emerged from the interface between cosmology and particle physics. Some emerged from pure observations — getting a better look at better data. The most pressing question for the Big Bang was a paradox hidden in the best evidence for its existence, the microwave background itself.

Using the CMB, astronomers could accurately measure the properties of the early universe in different regions of the sky. In particular, they could extract the temperature of the early cosmic gas with astonishing precision. To their surprise the temperature was the same no matter which direction they looked.[13] Even when their instruments were pointed in exactly opposite directions of the sky they found the gas temperature to be the same down to one part in ten thousand. This did not make sense. It was one thing for the early universe to be mostly smooth. It was quite another for it to be mostly smooth in exactly the same way everywhere throughout creation.

Why would cosmic conditions such as temperature and density be exactly the same everywhere? Since the universe began as a chaotic, expanding fireball of gas it would be reasonable to expect that some parts of the fireball ended up with slightly different conditions (e.g., temperature) than others. It was like reading the morning paper and finding out every single city on the planet had precisely the same weather.

Astronomers know how the universe expanded via Einstein's equations. The math makes it clear that regions of space now on opposite sides of the sky simply could not affect each other. In the parlance of the science, they could never have been causally connected. How then could conditions, deduced from the CMB, be the same on both sides of the sky? Without adding something to the story of the Big Bang these harmonious conditions appear to be an amazing cosmic coincidence. Astronomers hate that kind of thing. Amazing coincidences keep them up at night wondering what else is going on. If even light, the fastest thing in the universe, did not have time to make it from one region to the other, then how could the primordial gas filling these disconnected spaces end up with synchronized properties?

The data were speaking loud and clear. Opposite ends of the sky are too well matched to be explained with the standard Big Bang. This causal conundrum was the monster problem astronomers and physicists faced in the 1970s. If they could not solve it, their hopes for a rational scientific cosmology would collapse.

INFLATION GROWS ON ASTRONOMERS

In the history of human imaginings there are times when a single idea emerges to focus inquiry, bring clarity, and sow the seed for the birth of entire new universes. *Inflationary cosmology* is such an idea.[14] Inflation began as a bold attempt to solve the paradoxes and problems that plagued cosmologists at the end of the 1970s. The basic idea was simultaneously radical and elegant. With inflation, cosmologists imagined that *the part of the universe we can see* underwent a brief period of rapid expansion very early in its history. "Early" here is an understatement. The era of inflation begins when the universe is a mere 10^{-34} of a second old. That is less than a million billion, billion, billionth of a second after the Big Bang.

Inflation is the domain of head-spinning numbers. During inflation, the universe increased in size by a factor of 10^{43} (i.e., 1 followed by 43 zeros), and it all happened in just 10^{-33} of a second. During that sliver of time the cosmos grew in size from less than the diameter of a subatomic particle to the width of a softball. Then inflation shut down and the universe resumed the more leisurely expansion we see today. For comparison, in the last half of its life (the last 7 billion years) the universe has increased in size by only a factor of 1.6. Inflation was a crazy-brief period of expansion on steroids.

Inflation's brief period of hyperexpansion solved all the problems in standard cosmology. Most important was the causality issue. With inflation every part of the universe we see today *was* in causal contact before the hyper-stretching started. With a single change to the Big Bang (an early, brief period of hyperexpansion) the problems were resolved. It was hard for physicists and astronomers not to take notice.

Every good idea needs a champion. For inflation it was Alan Guth, a physicist at MIT. In 1981 Guth wrote a paper describing an inflationary model for cosmology. Some of his ideas had been proposed before, but Guth brought them together in a coherent, accessible way and added the catchy "name brand" *inflation*. What mattered most, however, was that Guth was not an astronomer. He was a particle physicist, and he built his inflation theory using tools from the empire of grand unification theories (GUTs).[15]

GUTs have been at the forefront of physics for decades. They are the attempt by theoretical physicists to understand the structure of matter and its interactions at the deepest level. There are four known forces in our universe: gravity, electromagnetism, the strong nuclear force, and the weak nuclear force. These forces are the only ways in which any of the "stuff" in the universe can interact. Particle physicists have long suspected that each of these forces was a different facet of a single "superforce." Unity from diversity is an idea with a powerful and mythic resonance. With grand unification theories, physicists predicted that if you could "heat up" the universe to ever higher energies, the different forces would each sequentially "melt" into the superforce just like individual ice crystals melt into liquid water. Heating up a tiny speck of the universe is exactly what particle physicists do in their giant accelerators, smashing subatomic particles together with tremendous energy.

Of course, the entire universe had already been through a stage when it was as hot as any particle physicist could ever want. Cosmology naturally gave physicists the GUTs laboratory they were looking for. Guth and those who followed built their inflationary cosmology using essential ideas taken from GUTs. The critical link between the particle physics and the cosmological physics was the energy source that drove inflation, something theorists called the "false vacuum."

While inflation may seem like the ultimate free lunch, taking a speck of the natal universe and blowing it up into everything we can see, there has to be a mechanism, a source of energy, to make it work. Drawing on ideas from grand unification, Guth and others imagined the early universe to be pervaded by a field of energy, a background, that mimicked empty space but had power to burn. This was the false vacuum. Particle physicists were used to this kind of quantum energy field from their explorations of matter and their work on GUTs. In their construction of inflation theory, Guth and others made the false vacuum field unstable, like a pencil standing on its point. This meant that sooner or later the false vacuum had to decay into a real vacuum and the energy in waiting latent in the field would be released. Inflation was the result of that decay. Kicking in just as the universe cooled to the right temperature, the energy released in the false vacuum's decay acted as a kind of antigravity afterburner. It repelled space-time and inflated a pinprick of space-time into the entire observable universe.

By building inflation from the framework of particle physics and grand unification, Guth and those who followed once again united the realms of the very small with the very large. Inflation has achieved its relevance by bringing the frontiers of cosmology and physics into a coherent and seemingly seamless whole. The effort saved the Big Bang and continues to bear fruit. But inflation's children have turned out to be far stranger and more numerous than its founders might ever have imagined.

MEET THE MULTIVERSE

In the years since Guth's first paper there have been many modifications to the idea of inflation. One line of inquiry looked deeply at its fundamental assumptions. From this work came an unorthodox understanding of the very concept of cosmology's subject matter. Of course, when it comes to considering the universe as a whole, orthodoxies are likely to be of little use.

Even a moment's consideration of inflationary cosmology tells us that there is a whole lot more cosmos out there than what we can see. After all, the basic tenet of inflation is to take a little corner of reality and blow it up into an entire universe. But what about the parts of the cosmos that did not inflate into our universe?

All inflation theories share the same characteristics. They all create a repulsive antigravity phase in the early universe and blow up a speck of space-time by a factor of about 10^{43}. In this way they all address the basic problem with Big Bang cosmology. But from their success comes the possibility of *eternal inflation*, an idea that raises the stakes on what we think of as the universe. Imagine a chunk of the universe undergoing inflation. Then imagine that only a small portion of this inflating space-time makes the complete transition from the false vacuum to the real vacuum. This portion will constitute a "pocket universe." It will expand at a normal rate, evolving galaxies and stars and planets and living, thinking creatures. Meanwhile all around it is a vast region of space-time still undergoing inflation. Why wouldn't other domains of this inflating cosmos undergo their own transitions to a real vacuum state?

Some physicists decided that these other regions were of interest and their calculations showed that a huge ensemble of pocket universes could be produced by inflation. Instead of one universe a "multiverse," an infinite universe of universes, emerges from the Big Bang. The multiverse is a dramatic reassessment of the very meaning of the word *cosmos*. It is one of the strangest ideas to emerge from modern physics, and it may also be inevitable. If scientists believe in some form of inflation, then it seems some form of eternal inflation must also occur.

OUR MEDIOCRE UNIVERSE

With all those extra pocket universes it would seem that physicists suddenly have a lot more to study. The multiverse, however, is not so transparent. The different universes would not be causally connected. No signals from one pocket universe could ever reach any other. There is, therefore, no way to see how they are constructed or to study them. In spite of this cosmic barrier there may be ways to tell something about the multiverse just by looking at our little corner of reality.

Physicists have long known that the particular character of physical law in our universe may be an accident of our history. Just as ice crystals will form in a seemingly random pattern, the exact way our four forces froze out of the superforce could have been different if cosmic history were run again. Gravity might have been stronger, or it might have been weaker, for example. The problem with changing something like the force of gravity or electromagnetism even a tiny bit is that all cosmic history shifts dramatically with respect to the creation of life. Weaker gravity can mean stars never form, which means no life. Stronger electromagnetism can mean biomolecules cannot form, and that obviously means life cannot form.[16]

For many years theorists have been bothered by this apparent fine-tuning of physical law with regard to allowing life to form. The "accident" of all four forces having exactly the right strength to make life feasible appears, once again, like an amazing coincidence. Intelligent design proponents have pounced on this conundrum as proof of fine-tuning by a creator. The concept of the multiverse may put scientists' worries to rest. If there are many, many universes, then there will be many, many different ways for the forces to freeze out. Fine-tuning is not needed at all. There will be many, many universes where life could never form, and there will be many, many universes that are amenable to life. In the census of "dead" and "living" universes we, not surprisingly, we find ourselves in the latter category. End of story.

Some researchers have even taken this reasoning further, arguing that since we would not expect to end up in a "special" universe (that would be more fine-tuning), then our conditions must be something like the average. Any randomly chosen universe must have a high probability of looking like ours, just as any randomly chosen middle-aged man on the street in the United States will be about five feet ten inches tall. In this view our mere existence becomes a powerful route to determine the statistics of pocket universes in the multiverse. Any universe with our properties must lie somewhere close to the mean or average of all the ones created in eternal inflation. It is a staggering idea even if we are a long way from providing any proof of its truth.

From the false vacuum to the multiverse, inflation has rewritten the story of cosmology. What began as a radical solution to a suite of paradoxes has grown into an ever-expanding paradigm for viewing what is and what can be. The story changes, but the underlying capacities of the narrative to force us out of our everyday reality does not. The Big Bang resonates with the myths of the Abrahamic tradition. It is something Christians, Jews, and Muslims might feel at home with. The new story of the multiverse echoes the infinite parallel universes of Vedic mythologies. The point is not which tradition the narrative of science recovers. As the physicist and writer Marcelo Gleiser has pointed out, there may be only a few choices in thinking about cosmology, and these choices have already been mapped out by mythology.[17] Instead we can see in cosmology the original aspiration to know the deepest truths of the world expressing itself in our investigations and our response to the telling of the tale.

COSMOLOGY AND MYTH

Science is facts; just as houses are made of stones, so is
science made of facts; but a pile of stones is not a house
and a collection of facts is not necessarily science.
      Jules-Henri Poincaré, *Science and Hypothesis*

"Your story sucks." My editor had a gift when it came to motivating
people. I was still new to popular science writing. It was just a side gig
next to my regular work as a researcher. I had been very lucky to get a
chance to write for *Discover*. The piece on stellar explosions I had just
finished was only my third for the respected magazine. It had taken
me a long time to write, and I took my editor's criticism hard. When I
asked him what was wrong, what exactly "sucked" about the piece, his
response changed more than just my writing. "There is no narrative
drive," he said, "It's just a bunch of facts strung together. No one will
get anything from it. No one will stay with it from beginning to end."
There was a pause on the line, and then he added emphatically, "There
has to be a reason for people to care about the story."

My editor was right. Science may begin with facts, but something
else has to be added. That is where stories come in. The stories of science emerge after the data are taken. They come from our desire for
relevance. The stories come because we care enough to pay close attention to those bare facts. We humans have an aspiration to know, to draw
closer to the source of our deepest experience of the world's beauty and
its power. The aspiration to understand the world and its ways drives
us to seek meaning in the data: pattern, form, structure. It is the route
from data to story, in the form of explanations, that reveals the roots of
science's capacity to make hierophanies and its roots in myth.

The data scientists collect with telescopes, microscopes, particle
accelerators, and PET scans are the raw material. The data are, in a real
sense, the world's voice spoken in the language of quantity. But there is
no meaning in the data by themselves. They are *just* numbers, and they
do not speak with a single voice but in a cacophony. To create mean-

ing, to find sense in the data, the scientist must be astute in listening for coherences and seeing patterns. When patterns are found they are pressed into the service of an explanation. Why is the pattern there? What hidden forces created the pattern? Most important, how do the specifics of the pattern and its explanation fit into broader explanations of the broadest issues we seek to understand: the history of the universe, the origin of life, the evolution of species?

The world gives us data. We look for patterns. Then we find a reason for the pattern, and that reason becomes a story. The stories cascade upward and are fit into bigger and broader narratives of our deepest, most compelling questions. This is where science touches the mythic. This is the chain through which "narrative drive" appears in the stories of science. The cosmology story I just told illuminates that chain, and myth is there in every link.

The cosmological narrative told in this chapter begins with data, but even there the mythic becomes apparent. Recall the story of the cosmic microwave background. My telling of the story emphasized *how* the data were discovered. There was the intercession of blind chance when Wilson and Penzias stumbled on the fossil signal in their microwave antenna. There was the demand for integrity in their tireless effort to root out the cause of microwave hiss. Finally there was the gift of clarity delivered in the realization that the signal was cosmological. Taken together, these elements of the story bring us face-to-face with the heroic aspect of science. It is a call to myth and a recapturing of mythic themes. The scientist is on the hero's journey.

Can the scientist be a mythic hero? It would be easy to dismiss the association simply as my decision to tell the story with this slant. But a glance through almost any account of science's progress, especially those intended for wide audiences, will reveal this quality in the telling. We cast the scientist in this role because the search for truth has always been a part of our mythic response to the world. It does not matter if the scientists really were heroic. They might have been self-serving schmucks. What matters is how the culture appropriates and tells the story to elevate it to a higher level. In a scientific culture scientists often play the role of shaman or hero in public narratives. We need them in this role because we have never lost our need for shamans and heroes to mediate our exploration of life's deepest questions. We cannot create resonant creation stories alone in our rooms. Cosmologies, or cosmological myths, develop as a societal project. In the past we required heroes and shamans to act as emissaries of these truths and as interpreters for our rapt attention, our awe. Now in a more democratic but no less mythical fashion we let scientists play the role.

What makes science particularly compelling is that its heroes can be any of us. We elevate some scientific characters, such as Einstein and Newton, to the truly mythic status of heroes because of their mysterious quality of "genius" (a word whose origins are linked to notions of spirit). Most scientists in the stories, however, can be average people. In public narratives scientists usually start off like everyone else. Nothing sets them apart aside from their desire to know and a tenacity to not let go of that desire.

Beyond the heroic story of "getting the data" our cosmological narrative also shows the transformation of observed pattern into narrative. For our story the pattern was found in Wilson and Penzias's microwaves. It revealed itself as the specific relation between the energy and wavelength of the microwaves. That pattern became meaningful through a relatively narrow narrative of primordial electrons finding primordial protons and bonding together into primordial atomic hydrogen. This very specific story then became embedded in a larger narrative of an expanding universe. In other words the narrative of the CMB was joined to the pattern Hubble found in his own study of galaxies. Together these were joined with the story of Big Bang nucleosynthesisis. By uniting all three stories, the broadest narrative of cosmic evolution emerged.

Myths are stories because that is how human beings create meaning. Stories are how we structure our response to experiences of the world's sacred character. Each step in our cosmology was, likewise, a story because narratives are still required in science. They create meaning from raw data: space and time expand; galaxies are pulled along with them; primordial elements form in a universe hot and dense enough for fusion reactions; CMB photons are freed when atomic hydrogen is first formed. Each story has its own narrative drive. This makes the story scientifically interesting and worthy of deeper study and articulation. Each story recalls the function of myth in explaining how things are and why they came to be. Once the highest level of the narrative is reached, however, the call to the mythic stands fully revealed.

The narrative of modern cosmology, the Big Bang, is a myth because we respond to it as one. It calls us to an attentive hearing. This response was taken on by uncounted generations before us when they stood before their own narratives of the cosmos. To position ourselves ready to recount the "true" history of creation is to adopt a stance our Paleolithic ancestors were familiar with. We can imagine the shaman standing before the opening of the shelter. Behind her the darkness of the night sky opens. Her people feel the crushing sense of awe, and she interprets that feeling for them, deepening it with their own collective story of creation.

Like our ancestors we hear science's cosmological narrative and stand ready to be bought out of the profane world. When we decide to make the effort to hear, to read, to listen, or to watch we bring ourselves into the posture of receptivity. It is an ancient calling that science now mediates and addresses. There are many ways the story comes to us: the *NOVA* program on the Big Bang, with its surging music; the *Newsweek* cover story on the early universe, with its vivid descriptions; the visiting scientist's public lecture, with dramatic Hubble space telescope images of young galaxies. When we are young we are taught the stories in school. When we are adults we cross paths with the stories by chance or by choice. Even if we do not recognize them the stories appear, as background in movies, novels, and music. At some point we may finally stop, listen, and reflect. In those moments we allow the ancient sense of awe to have its say. Through the stories of science that sense of awe, of the sacred, becomes realized, deepened, and interpreted yet again. This has always been the function and the power of myth.

Who in their day-to-day activities thinks about the essential nature of space and time? During the daily commute or when making dinner, there is no sense that space and time are special and need to be noticed. Yet in our telling of cosmology, space and time were principal actors. They were something fundamental, needing consideration on their own. This is another, critical mythological function of science. It brings the character of sacredness out of the profane. The idea of space-time, a fruit of scientific practice, is a hierophany. It is a gateway to the sacred because it brings the background of experience to the fore and forces us to consider what had been invisible. On hearing the description of Einstein's work and the possibilities of warped 4-D space-time, we pause. For a moment we step outside ourselves and the world is illuminated, made more than just the stage of our personal concerns.

Here Mythos appears in Logos. The space-time we learn about in the story of cosmology evokes a sense of the world's great wonder, of its possibilities manifest as an unseen but present reality. It evokes a sense of the world's sacred character in its abiding and essential mystery. We encounter other hierophanies in cosmology: the four basic forces splitting off from the unitary superforce; the quantum ripples surging through time and becoming structures such as galaxies, which lead to stars and planets; the dark matter pervading all creation like an invisible ocean. When apprehended with a receptive ear, each idea is an actor in the cosmological narrative that can transport us away from the everyday to reveal a living world of wonder always present, always at work.

Science has its roots in myth because like myth it opens up hierophanies. The vast ideas of space-time and grand unified theories have a power that goes beyond their ability to explain tables of numbers in scientific papers. The ideas themselves are mythic because they connect us to millennia of aspiration. They contain the element of myth that has been constant across thousands of human generations. These ideas, embedded in their narratives, drive open a sense of the world's sacred character. They can make us more alive to the act of being alive. They create a clearing in the daily tumble of our lives. There we can find the one, true location where science and spiritual endeavor, science and religion, come into harmony.

Science recovers the themes and attitudes of myth in many ways and in many settings. Cosmology was the first and most obvious domain where the intersection of science and myth became apparent. We do not need to travel to such distant environments to see the braiding of their workings, however. Even here on Earth, in the call of storm and flood, we can find science emerging from myth and the recollection of myth within science.

CHAPTER 7

The Deluge This Time

Climate Change and Flood Myths

A flood will sweep over the cult-centers;
To destroy the seed of mankind . . .
It is the decision, the word of the assembly of the gods.
Sumerian Epic of Atrahasis

Everybody talks about the weather but nobody ever does
anything about it.
Charles Dudley Warner

God gave Noah the rainbow sign
Said, "No more water, but fire next time"
Pharaoh's army got drownded
O Mary don't you weep
African American spiritual,
interpreted by Bruce Springsteen

It began as a labor dispute of cosmic proportions. For eons the mighty high gods of Babylon had forced the lesser deities into service. Anu was the god of heaven, Enhil the god of the Earth, and Enki the god of the freshwater oceans. Each ruled his domain as an absolute lord with power over the lower gods. Enhil was a builder. He had grand plans for the Earth: vast cities and farmlands. For thousands of years he had pressed the lower gods into labor, endlessly building lifelines on the land, digging canals, and dredging the Tigris and the Euphrates Rivers. It was ceaseless, backbreaking work even for a god. Finally they had had enough and a great bonfire was lit. The gods threw their tools on the pyre and marched to overthrow Enhil.[1]

Enhil startles awake as his vizier Nuskul urgently shakes him to consciousness. "My lord, an army of lesser gods comes arrayed against you," he says. Enhil walks to the window and sees the angry mob outside his temple. In spite of his power Enhil grows pale with fear. Even the great deities must contend with politics and its deadly consequences. "Call the other great gods," Nuskul tells Enhil. "They are wise and can offer you guidance." Anu and Enki soon arrive resplendent in their robes and luminous in their power. Together they take a grim survey of the host of enraged lower deities outside. "Find the ringleader," Anu snaps. "Make a sacrifice of him, and you can quell this rebellion." Nuskul steps outside. "Tell me," he calls, "who is your leader? Who is it that speaks for you?" The lesser deities are not fools. They call back with a single voice, "Every single one of us has declared war today." Like thunder their righteous anger and determination echo across the sky. The depth of the crisis is now clear to Anu and Enki. "You have worked them far too hard, Enhil," they say. "You must find a better way."

Thus man is born. The great gods decide a sacrifice must still be made, but this time it will serve a greater need. The womb goddess is summoned and a single lower god is slaughtered. The blood of Gesthu-e, a god whose name means "wisdom," flows across the altar floor. The womb goddess gathers his life fluid and carefully mixes it with clay. From this dark substance the race of men is shaped.

Fired like bricks in a kiln, human beings are given life and purpose. They replace the lesser gods as workers digging Enhil's grand canals. For 1,200 years Enhil's labor requirements are satisfied. But the humans grow in number, and the Earth god finds himself besieged by the din of their activity. Like a bellowing bull, the humans fill the land with the clamor of their ceaseless mating, arguing, and fighting. Soon Enhil can find no rest. Deprived of sleep, he calls again upon the other gods. "I will send down a plague to wipe them out," he cries. The gods reluctantly agree, and the plague god sweeps down upon the people.

Death winnows the human numbers. Fathers, mothers, and children are taken quickly and without pity. In their anguish the people call out to their wise king Atrahasis for help. The king in turn beseeches Enki, god of waters, for salvation. In his compassion Enki relents and tells Atrahasis that humans can save themselves by restricting their prayers only to the plague god, shaming him into ending the great sickness. After months of determined prayers and sacrifices the plague god finally capitulates in embarrassment. With a wave of his hand the sickness is lifted.

When Enhil discovers Enki's betrayal he is enraged. He tries again to wipe out the human race, first using drought and then famine. Each time Enki foils his plans. Finally he has had enough. After obtaining consent from the other gods Enhil plans a flood that will drown the world of men once and for all. The other gods have also grown tired of humans and promise their collaboration. Enki too is forced to assist. But he is fond of King Atrahasis. He breaks ranks with his fellow deities. Whispering to Atrahasis through the walls of his reed house, Enki instructs him to build a boat and save himself, his family, and all Earth's creatures.

When the storm finally comes it shakes the world to its foundations. Even the gods tremble in fear at the forces they have let loose. Like a wild ass the winds bray and howl. The darkness that comes is total. The sun is swept back behind black clouds and is lost, it seems, forever. As the storm ravages the land the great mother goddess cries in despair, weeping for the dead humans who clog the rivers like dragonflies. "It has been too much!" she wails, blaming the great gods for their failure to manage the world's affairs.

Atrahasis and his family try to ride out the terrible storm in their simple reed boat. The waters rise, and they watch in horror as friends and companions are swept away. The endless nights of rain and wind press down, and they cower before towering waves that surge like great beasts. The howling storm is relentless. They fall, battered, on the deck of their frail craft.

After seven days the storm finally breaks. The winds slowly die and the pale sun can be seen through the screen of clouds. The floodwaters begin their slow, quiet retreat. In the growing stillness the reed boat comes to rest. Atrahasis stumbles to the ground dazed by the ordeal and the reality that he and his family have survived while all others have disappeared. In thankfulness for his rescue the king makes a sacrifice on a rocky outcrop surrounded by brown mud and debris.

The gods are drawn like flies to the fragrance of Atrahasis's burnt offering. They quickly find the king and his small band of survivors. Once again Enhil, in spite of his lashing by the mother goddess, is furious at Enki's betrayal. The god of waters remains unbowed. "I made sure life was preserved," Enki retorts sternly. "That is what matters most." Now, after so much terror among both gods and men, Enhil relents. He is ready to compromise. Enki convinces him to muster his compassion and find a new solution to the human problem. Never again will Enki allow the rising waters to threaten the world.

. . .

We must have seen devastation like this many times. Each flood, each catastrophe, must have left a mark in our minds. The narrative of a great deluge appears in almost every human culture at every epoch of prehistory. The myth of Atrahasis is one of three flood stories in the Babylonian tradition (some argue it is also the precursor of the biblical flood myth of Noah).[2] Other cultures have other forms of the story. In Sumatra there is the story of Debata, the creator god, who floods the world when it grows old and dirty. In India there is the myth of Manu, the first human, who is saved from the world-destroying flood by a great fish he spared when it was young. In Australia there is the myth of the creator Bunjil, who grows angry with the people for their evil and floods the world by urinating in the oceans. While the specifics of the narrative differ, the central themes of decay, vengeance, and renewal appear many times. There are more than three hundred variations of this basic narrative documented. The righteous flood, it seems, is a ubiquitous human story.

In some cases evidence exists for a historical event at the source of the myth. There has been a small industry of workers attempting to link the biblical account of a great flood with rapid glacial melting, a tsunami, or some other "scientific" explanation. Turning from the Bible to the general prevalence of global flood myths, writers have speculatively connected a great flood with everything from rising oceans at the end of the last ice age to an asteroid impact.[3] While it may or may not be possible to connect a specific flood myth of a specific people with a specific geologic event, the attempt is beside the point for the questions we are asking here. Given the propensity of human populations to live in coastal regions and river valleys, the experience of storm and flood must have been common. In our attempt to follow science's living roots in myth the flood narrative illustrates something remarkable that must be explored more deeply. Flood myths can perhaps reveal how experience is transformed into a narrative rich with meaning. That narrative is elevated to myth when it transcends the specific to touch the sacred character of experience. This elevation holds true even when that experience is one of terror and loss. The desire for explanation is never lost but the story rises beyond its explanation to become something richer and more fertile.

Changes in climate and the corresponding dramatic variations in water levels and storm activity can occur on relatively short time scales (hundreds, thousands, or tens of thousands of years). Our genus *Homo* has seen many of these cycles as glaciers advance and retreat and land bridges between continents emerge and then are resubmerged. Our own species *Homo sapiens* has certainly been around long enough to see some of these changes. Myth likely represents some cultural memory that remains of these experiences. The transformation of myth into sacred narrative endows memories with moral relevance in a context of broader, deeper themes. This is what makes flood myths so powerful. In the narratives of deluge we see experience of the natural world transformed directly into thematic, sacred stories.

Now we are on the edge of our own experience of rising water and storm. The narratives of a great flood have suddenly gained relevance that extends beyond academic discussions of myth. The imperatives of global warming bear down on us, and once again we will face the overlap of the mythic and the scientific. Climate science is another arena in which echoes of our long heritage of sacred narratives can be heard. In the science of climate change we will find a different facet of the interweaving of science and myth. The complementarities we uncover will illuminate the transition from data to narrative to sacred story once again, which will help us to understand the central theme of science and spiritual endeavor. In this exploration, however, we will encounter a more troubling possibility. The pairing of flood narratives and climate science may also present us with the shadow of future dramas we have yet to enact.

DISCOVERING RAPID CLIMATE CHANGE

Camp Century was no place for absentminded professors.[4] It was a military research base located one hundred miles out on the ice sheet that covers Greenland. Camp Century was a small town built under ice. With winds reaching to 125 miles per hour and temperatures dropping to as low as -70 degrees Fahrenheit, constructing the installation below the snowy surface was, without doubt, a good idea. Up to two hundred men lived and worked below the surface. They moved from room to room through ice tunnels that could stretch as far as three football fields. To power the installation the air force dragged a portable nuclear power plant across the ice.[5] The construction and maintenance of Camp Century in the early 1960s was an engineering nightmare that pushed men and machines to their limits. Limits were, however, exactly what the military was interested in. This was the Cold War, and the United

States was determined to use its power to push the frontiers of science to new extremes when it served the national interest. The history of warfare is full of military campaigns done in by weather. Funding for climate science was wide open.

In the early 1960s the study of climate, or long-term weather patterns, was still in its infancy. One of the biggest problems the new science faced was lack of data on climate change over thousands of years. Camp Century was ready to solve that problem. The scientists who lived and worked in its frozen hallways knew there were climate secrets buried far below them in the ice they stood on. From 1960 to 1966 Camp Century was the home of a radical and radically difficult experiment to map out the history of the Earth's climate.

The U.S. Army's Cold Regions Research and Engineering Laboratory (CRREL) had its main ice core drilling operation at Camp Century. In a freezing hall carved out of the snow pack, the Danish scientist Willi Dansgaard and his American counterpart, Chester Langway,[6] led an international team attempting to drill into the ancient Greenland ice. Greenland has been very cold for a very long time. Most of the landmass is covered by an ice sheet that can reach thicknesses of two miles. Its massive crust of ice is maintained by annual snowfalls that are packed one on top of the other. As years turned to centuries that turned into millennia, the strata of ice became a kind of frozen layer cake, with each layer comprising a record of that year's climate. The CRREL mission was to drill downward at Camp Century and retrieve long cylindrical sections of ancient ice — ice cores — from deep below the surface. Within each layer of ice was a chemical marker that could tell scientists the average temperature in Greenland the year the layer was deposited. The temperature marker consisted of two different forms of the element oxygen, called isotopes. The relative abundance of the oxygen isotopes made them very sensitive to atmospheric temperature, and they served as a proxy thermometer. By analyzing the isotopic content of the different layers of ice, Dansgaard hoped to map the Earth's climate back to the end of the last ice age.

Ice ages were still a mystery in 1960. It was not clear how they started, how they ended, how many of them the Earth had passed through, or how long they lasted. Ice ages *had* happened; that much was clear. Over the past few million years mile-thick oceans of ice had covered vast stretches of the northern hemisphere, grinding south and then retreating at least four times. Between these ice ages the Earth remained warm for thousands of years. The entirety of human history and civilization from the Neolithic period to the development of the first cities had occurred during the current interglacial period. The grand challenge of paleoclimate studies was to understand the dynamics of these ice ages. What caused them, and what caused their timing?

In the 1960s paleoclimate was a matter of purely academic interest. Few scientists then were willing to consider climate change on time scales that could affect their own lives. The consensus among scientists was that any profound shift in global climate, the kind of thing that raised oceans, melted continent-sized glaciers, or changed mean temperature by whole degrees must take many thousands of years. If there was change coming it was so far in the future that no one needed to worry. That scientific consensus was, however, nothing more than prejudice. In lieu of hard data and the difficult analysis that would follow, scientists were left with hunches and their innate biases. Most of these would turn out to be wrong. A storm was building, but few could see it coming.

The work at Camp Century was grueling. Drilling occurred in segments. The drill bit would move down a few meters, the core would be extracted, and new piping would be added. The bit frequently got caught and needed to be repaired or changed. When the cores were taken they had to be treated with extreme care so as not to damage the integrity of the ice. All this had to be carried out under freezing conditions on the planet's most inhospitable terrain. After six years of effort Dansgaard, Langway, and the Camp Century team prevailed. Together they finally managed to drill all the way down to bedrock 1,300 meters below the ice. Their ice cores produced a continuous record of our entire interglacial period, all the way back into the last ice age.

Dansgaard was responsible for the analysis of the ice cores. Like the rest of the paleoclimate community, he was interested in mapping what he expected to be long, slow variations in climate that comprised the ice age cycles. In the 1950s oceanographers had done their own drilling, bringing up layers of silt and clay from the ocean floor. These studies showed evidence only of the expected long gradual changes in climate. A few researchers criticized these studies, claiming that "slumping" of sea clay and mixing of the sea floor layers by burrowing worms made more rapid climate changes invisible to the oceanographers' methods. For the most part the criticisms were ignored. The long time scales found by the ocean floor studies matched a prevailing bias in geological sciences termed "gradualism." The gradualist perspective claimed that all processes shaping the planet occurred slowly. Many small, successive changes over time created the different eras recorded in paleontology. The dominance of gradualism made many scientists inclined to focus on data that matched its preconceptions.

Still, there were a few lines of evidence that pointed to rapid climate change. Using hardy pollen spores that could survive encased in layers of undisturbed peat and mud, Scandinavian scientists had concluded that rapid and dramatic oscillations in temperature had occurred in Earth's "recent" climate history. In their examination of changes in spore types found in deep layers of ancient northern lakes they found evidence that something violent shook the climate some twelve thousand years ago, just as the last ice age was ending. The spore record showed a period of quick warming, with temperatures higher than those of the modern era, followed by sharp and sudden cooling — a return to frigid ice age–like tundra for many hundreds of years. This period was named the Younger Dryas, in honor of the tundra flower *Dryas octopetala*, whose pollen made the discovery possible. Although the Scandinavian evidence was impossible to dispute, most scientists dismissed it as a local event. They were not willing to believe the entire world could undergo such rapid climate oscillations. Daansgard's Camp Century data forced them to reevaluate this perspective.

It took a long time to complete the isotopic analysis of the Camp Century ice cores.[7] Each layer had to be carefully separated and samples were run through mass spectrometers that separated the different oxygen isotopes. When the temperature record was complete the disparity between data and prejudice was as clear as the dark line on the graph. Dansgaard and his colleagues saw a long period of roughly constant temperature stretching back 8,000 years, the era of modern human civilizations, preceded by a gradual cooling. The ice layers captured the transition from the warmth of our current climate to the frozen glacial age 14,000 years ago.

Superimposed on top of the smooth transition from the warm interglacial period to the frigid glacial era were a series of "spectacular" short-term shifts. The Camp Century data showed powerful oscillations from warm to cold climate occurring about 12,000 years ago. This was approximately the period and duration of the Younger Dryas. Dansgaard's analysis of the Camp Century data bolstered the spore data. Together they provided strong evidence that the Earth's climate could undergo change on time scales far less than the multi-millennial epochs most climatologists expected. It was a milestone that set scientific opinion on a new course.

As the decades passed climate change passed from an academic specialty to the realm of public policy. The prospect of abrupt shifts in global weather patterns became a new and frightening possibility. To finish the job started at Camp Century, new ice core drilling projects were initiated around the globe. One of the most ambitious of these focused once again on the Greenland ice sheet. In the mid-1980s U.S. and European teams used new drilling technology to mount an offensive on the summit of the Greenland ice plateau. The plateau constituted the top of the entire Greenland ice sheet, and it was situated so high that scientists and engineers working there had to contend with altitude sickness among their other worries. Stationed at two sites, the teams drilled more than 3,000 meters down to bedrock. Each team recovered pristine cores that provided a high-resolution chronology of 110,000 years of climate, including the last interglacial-to-glacial cycle. The story told in the ancient ice was compelling and frightening.

Once again Willi Dansgaard led the European group. Their analysis of the plateau ice cores, confirmed by American data from 23 kilometers away, showed powerful, rapid shifts in climate. Many times over the past hundred millennia Greenland's seasons were shifted up or down by 7 degrees in just fifty years. Comparable changes in the modern era would mean devastation of human civilizations. One American researcher recalls a moment of truth in 1992 when a mere visual inspection of an ice core made the speed of climate change apparent. The young researcher could see obvious differences in the ice layers at the onset of the Younger Dryas period. Recorded in darkening layers was a visceral confirmation of global climate changing in just three years. Now there could be no doubt. Climate change could come hard and fast.

GREENHOUSE WARMING UNVEILED

The urgency of climate science shifted as the world moved from the Cold War era of Camp Century to the Greenland ice plateau drilling. In the space of those three decades scientists and the public slowly came to recognize that climate change was more than an academic subject restricted to ancient epochs of geology. Sometime around the late 1980s a new term was added to the lexicon of popular consciousness: the greenhouse effect.

The greenhouse effect was not discovered in the 1980s. As early as 1859 Joseph Tyndall had recognized that atmospheric gases were raising the planet's temperature by least 100 degrees. Planets such as Mercury with no atmosphere absorb the energy in sunlight and warm up. At the same time their surfaces radiate some of this heat back into space. Eventually a balance is achieved and the airless planet reaches a stable temperature. The Earth is hotter than this balance would suggest. On our planet the atmosphere was somehow trapping a fraction of the heat radiation reemitted by the Earth's surface and raising the temperature of the entire planet's surface. This radiation trapping is exactly what the glass in a greenhouse does to maintain balmy temperatures for plants on even the coldest winter days. Something in the chemistry of the atmosphere was acting like glass in a greenhouse, keeping absorbed solar energy from being reemitted into space. In a series of experiments Tyndall found the "greenhouse gases." He showed that while water vapor was most important, carbon dioxide (CO_2) was also a powerful energy absorber.

Fifty years later the Swedish physicist Svante Arrhenius was, like so many other scientists, seeking an answer to the riddle of the ice ages. His research was focused on CO_2 and ice ages, but instead he stumbled upon a remarkable link between human activity and planetary temperatures. After a painstaking calculation Arrhenius found that decreasing atmospheric CO_2 concentrations by just one-half would lower global temperatures by 4 or 5 degrees Centigrade. This kind of cooling must accompany the onset of an ice age. He and a colleague then decided to find what processes could cause large-scale changes in atmospheric CO_2. To their surprise they found that fossil fuel consumption — coal in the days of Arrhenius — was a significant factor. Human activity was adding CO_2 to the atmosphere at the same rate that natural processes were cycling the gas. People were altering the CO_2 budget of the entire planet. Arrhenius's calculations showed that the extra CO_2 could warm the planet over a time scale of many centuries. He considered this a good thing. What Swede would argue with warmer winters? The link between CO_2 and planetary warming had been established, but at the turn of the twentieth century no one saw any reason to worry.

As the twentieth century matured the fledgling science of climate was not inclined to take up the issue of CO_2 greenhouse warming. The focus remained on the mystery of the ice ages. Where interest in

CO_2 did exist it tended to focus on how the gas moved between the atmosphere and the oceans. It had already been clear to Arrhenius that the ocean would provide a massive "sink" for CO_2. It was expected that most of the extra carbon dioxide emitted by humans could easily be dissolved in billions of tons of ocean water. It was not until the 1950s that the first hints of the ocean's real appetite for CO_2 were pinned down.

In 1954 Roger Revelle of the Scripps Institution of Oceanography began an extensive project to explore the flux of CO_2 into and out of the oceans.[8] Using his expertise in ocean chemistry, Revelle soon discovered a series of chemical reactions that could brake seawater's capacity to absorb greenhouse gas. He concluded that previous calculations had been off by 90 percent in terms of the ocean's uptake of carbon dioxide. The ocean, it seemed, would have a hard time digesting much new human-released gas. Revelle's result was a milestone in climatology because it showed for the first time that CO_2 pumped into the atmosphere was likely to stay there.

A monitoring station set up high on the Hawaiian volcano of Mauna Kea soon confirmed the calculation. The curve of atmospheric CO_2 concentration taken from the Mauna Kea station showed the whole planet breathing. The CO_2 rose and fell as northern hemisphere plant life grew and died off. But each peak was higher than the year before. The CO_2 in the atmosphere was increasing at a rate that was close to Revelle's prediction. It was becoming clear that the oceans would not help us. If we continued to increase our industrial CO_2 production, then the concentration of critical greenhouse gases would increase as well. Revelle saw what might be coming when, in 1957, he wrote, "Human beings are now carrying out a large-scale geophysical experiment of a kind that could not have happened in the past nor be reproduced in the future."[9]

In the years that followed some scientists began to speak of the dangers humanity might face from a runaway greenhouse effect. But in spite of their hopes for simple cause-and-effect relationships their science got bogged down. The Earth's climate system was intrinsically and sensitively interconnected. The atmosphere responded to the oceans, which responded to the land, which was shaped by biological activity, which was determined by the atmosphere. It was a mess. Sorting out the web of relationships required years of work, massive international data collection efforts, and a willingness to create new ways of looking at complex systems. Much of the 1970s and 1980s were spent clarifying how CO_2 cycles through the climate system and building a global census of the gas. At the same time scientists were hard at work establishing the link between CO_2 and global temperature in Earth's climate history. Without this step, articulating possible futures would remain a dubious endeavor. Once again the world's ancient ice held the keys to understanding.

The cores drilled at Camp Century had been analyzed by melting and chemically analyzing the ice. By the 1980s researchers had found other, more sophisticated ways to extract information from the cores. Within each core were tiny bubbles of air trapped when the snows fell five thousand, ten thousand, or one hundred thousand years ago. The CO_2 in these ancient air samples remained intact and at the same concentrations as when the ice formed. The trick was to extract the air bubbles without contaminating the samples. After long years of work scientists finally figured out how to sample the CO_2 content in these minuscule samples of paleoatmosphere. By 1985 French and Soviet researchers in Antarctica had extracted a 150,000-year record of atmospheric CO_2 concentrations. It did not require a Ph.D. in climate science to draw the conclusion that some kind of connection between CO_2 and temperature existed. With every rise in global temperature there was also a rise in global atmospheric CO_2. With every drop in global temperature there was a drop in CO_2. While it was not always clear which came first, the rise in temperature or the rise in carbon dioxide, there could be no doubt that a correlation existed between global climate change and CO_2. The link did not bode well with scientists who saw human activity already pushing CO_2 levels far beyond anything that had existed in the past 150 millennia. If CO_2 meant more heat in the atmosphere, then it would appear that we were setting the stage for a global sauna.

PREDICTING CLIMATE FUTURES

What would happen as CO_2 and other potent greenhouse gas levels rose? As the greenhouse effect began to seep into the collective consciousness this question became central to the debate about what, if anything, human beings should do to moderate CO_2 levels. Arrhenius, after all, had thought a little warming would be a good thing. Should the world simply invest in more air conditioners, or was the end of civilization the only logical outcome of the global warming? Throughout the 1980s and 1990s climate modelers, scientists who used massive computers to simulate the evolution of long-term weather patterns, struggled to extract predictions they could stand by. Their effort pushed climatology straight to the front of the rapidly moving frontier of high-performance supercomputers.

Throughout the middle decades of the twentieth century scientists had attempted to build simple mathematical models of the climate. These pencil-and-paper calculations had proven useful in articulating basic features of the climate system, such as simplified relationships between CO_2 and temperature. But as the web of forces affecting climate were mapped out it became clear that the system was far too complex to be treated with simple pencil-and-paper methods.

In a happy coincidence the need to understand climate's staggering complexity emerged alongside the one tool that could manage it. The digital computer had been born in the years following World War II. Its speed and power had grown steadily during the first years of its operation. Climate scientists were quick to use this tool in their studies. By the 1980s, however, exponential increases in speed and memory ushered in the era of supercomputers. With these ultra-powerful machines researchers could take a real stab at modeling Earth's climate system in its full intricacy.

Computer simulations solve the basic equations governing atmospheric physics and chemistry on a virtual Earth broken up into grid points. A map of the Earth is covered with squares. Then the equations are solved at the center of each square and the climate on the entire virtual planet is advanced in baby steps. The computer programs scientists developed for this task were called global circulation models, or GCMs.[10] These were large, complex computer codes designed to predict the properties of the atmosphere: its circulation patterns, moisture content, temperature, and so on. Computer models of the ocean were also being developed at this time. By the 1980s computers were just becoming powerful enough to try linking ocean simulations with atmospheric simulations. This was a crucial step. Ocean and atmosphere models had to talk to each other because their strong feedbacks ruled global climate systems. As the decade progressed a few teams around the globe struggled to push their simulations to the point where reasonable results would emerge. Even with the best supercomputers of the day these models were crude. It was slow going, requiring long hours of coordinated effort by many researchers around the world. Success did not come quickly or easily.

To make a GCM work, a truckload of assumptions had to be built into the code. The machines were still too sluggish to track all the different processes at work in the Earth's climate. At first the results from different codes gave inconsistent answers. If the various GCMs built by various groups all gave varying answers to the same question it meant trouble. How could scientists know which code to trust? To remedy the situation researchers set a standard using the simple question of how much temperature change comes from doubling CO_2 as a test. When the models gave different answers scientists had to review how poorly understood processes had been approximated in different codes. Many aspects of climate science were still murky. No one was quite sure how clouds and their reflection of sunlight played into driving climate. The physics of clouds and other factors acting in the GCMs had to be worked out. An international effort to hunt down the uncertainties and develop ways to properly deal with them continued throughout the 1980s and 1990s.

Scientists were aided in their efforts by stunning advances in super-computer technologies. In 1985 the resolution of the models was poor at best. Each little square, or pixel, in the Earth-covering grid spanned one thousand miles on a side. It was like using a digital camera with just one hundred pixels to take a family snapshot. A lot of physics was lost in those giant, unresolved stretches of planet. By 2002 the best climate models had zoomed in with a resolution of less than a few miles. Also, by this time fully merged ocean and atmosphere models were the norm. The meaning of GCM changed from "global circulation models," which focused only on the atmosphere, to "general climate models," which could swallow the whole Earth system enchilada.

The new models were put through their paces. First all the GCMs were able to recover well-known aspects of atmospheric circulation patterns such as the jet stream and the large-scale eddies that bring warm tropical air up to the poles. Then, as a test, the models were set the task of nailing details of Earth's climate history. Using data from the ice cores and other sources, scientists already knew a lot about the climate fourteen thousand years ago. With these data as input conditions the new GCMs could simulate important characteristics of ancient climate epochs. This was an important hurdle for the models.

By the end of the 1990s all the major GCMs gave similar answers to the most important problem — the climatic response to increases in CO_2. The models were uniform in their predictions, even though the precise values of temperature rise varied. Many scientists increasingly felt they could now bracket the possible effects of global warming. In other words the models did not all give the same answer, but a consensus was growing that they were giving a good idea of the range of heating that could be expected. In these models doubling CO_2 levels leads to an increase in globally averaged temperatures of between 1.5 to 6.0 degrees. Skeptics of global warming had always criticized the models as unreliable and contradictory. Now they were faced with a dilemma, as every GCM worthy of the name showed that global warming was the *inevitable* consequence of CO_2 increases. The CO_2 greenhouse effect was robust. There was no way out.

Increasing the temperature a few degrees may not seem like much of a problem until you consider that these are globally averaged temperatures. Raising the temperature of the entire planet by 3 degrees requires a hell of a lot of energy. All that energy "suddenly" dumped into the climate system was the real problem. In many ways the term *global warming* is less descriptive of the dangers facing humanity than the alternative term *climate change*. Dump a planet-load of energy into the strongly interdependent climate system of oceans, atmosphere, biosphere, and so on, and something has to give. By 2000 climate scientists understood that what humanity was likely to face was a climate that did not uniformly get hotter but one that was uniformly more energetic and unstable.

It is important to recognize what the models can and cannot do. A world with 1.5 degrees of warming may be uncomfortable, but it also may be manageable. A world with 6 degrees of warming could be one that human civilization, in its current form, would find very challenging indeed. The models were showing us possible futures. Also, the predictions of the models on smaller regional scales were not uniform. Different GCMs predicted different consequences of greenhouse warming for different regions of the planet. Taken together, however, the models that predicted the greatest changes led to a frightening picture of a world shaken by a rapidly shifting climate. If the worst predictions come true, then humanity will face stronger storms and dramatic, unsteady shifts in weather patterns. Frosts will come later, and rainfall will become more uncertain. The consequences of these changes may be hard for both agriculture and ecosystems. The oceans may become more acidic, threatening delicate marine life communities. Tropical disease may spread in regions of strongest warming, as would invasive species such as weeds. Most troubling in these models was the prediction of rising oceans. Strong global climate change, it seemed, invariably meant global flooding.[11]

OCEANS, ICE, AND FLOOD

Glaciologists are used to flooding on grand scales and have a name for this phenomenon — jokulhaup. In the late 1940s geologists recognized that the vast ridges that shaped the eastern badlands of Washington State had been formed in titanic jokulhaups at the end of the last ice age. Hundreds of square miles had been reshaped in just a few days as a fresh-water ocean broke through its glacial dam and ripped across the land. Floods, ice, and changing climate have always gone hand in hand. The greenhouse effect, driven by human activity, might be no different.

From the GCMs, and other methods, scientists could see that rising sea levels and flooding would be driven by a variety of forces. Simply adding so much heat to the world's oceans will make them expand and push sea levels higher. In the twentieth century sea levels had been measured to increase 10 to 20 centimeters. Such a change represented a tenfold increase over the last millennium. By the beginning of 2000 scientists were certain that this rise in sea level was in part due to thermal expansion. The increased storm activity predicted to come from climate change would also produce flooding because storm surges — the local increase in sea level as storms push water before them — would become more frequent and more intense. During this time scientists also came to recognize that a confluence of processes affecting sea levels might combine to drive the most frightening of all climate change predictions.

The ice cores scientists had drilled so assiduously showed them that climate could change rapidly. The powerful general climate models they constructed over decades showed them that the planet was, on average, bound to grow warmer as energy continued to be trapped. Finally, con-

188 / The Terrain

tinuing research on glacial mountain and polar regions revealed these areas' sensitivity to changes in climate systems.[12] The world's ice sheets held tremendous storehouses of water. Contained within them was the potential for a disaster of biblical proportions.

The role of glacial melting from increased global temperatures was more uncertain than other causes of sea level rise. Certainly evidence was growing that melting was under way. Throughout the 1990s data accumulated that showed mountain glaciers disappearing at an alarming and unprecedented rate. In the Arctic researchers and the U.S. Navy (using its fleet of nuclear submarines) found a continuous decrease in the thickness of the polar ocean ice cap. Entirely open, ice-free summertime arctic oceans appeared to be a growing possibility. Beyond all of these effects was the disturbing uncertainty associated with the vast ice sheets covering Greenland and Antarctica.

More than a trillion cubic tons of ice lies locked up in the West Antarctic Ice Sheet (WAIS). This enormous mass of ice holds far less water than the rest of the Antarctic, but it is separated from the rest of the continent by a range of high mountains. In 1968 the maverick glaciologist John Mercer had noted that the entire mass was held in place by floating ice packs, called shelves, at the head of the WAIS.[13] Glaciologists had long studied the phenomenon of glacial surges, which occur when the movement of ice suddenly speeds up, dumping skyscraper-sized blocks of frozen water into the oceans. Mercer claimed that even small changes in temperature could make the ice shelves unstable, allowing the WAIS to surge forward and break apart. At the time most scientists dismissed Mercer's ideas. As *global warming* became a household word researchers looked at Mercer's ideas again, focusing attention on the possibility of catastrophic surges in both the Greenland and Antarctic ice sheets. Studies of ocean levels at the end of the last ice age showed sea levels had risen by 16 meters in just a few centuries. The cause of such a rapid rise seemed to be something happening in Greenland and the Antarctic.

In an effort to understand the dynamics shaping the great ice sheets scientists mounted ambitious efforts in the 1990s to quantify their properties and movements. The physics of glaciers and ice sheets depends on many factors. Detailed mathematical models of ice sheet formation offered some consolation, showing long time scales for their formation and destruction. These ice sheet models, unfortunately, had huge uncertainties. Empirical evidence taken directly on the ice and from the sky painted a more troubling picture to researchers. Satellite images showed that smaller Antarctic ice shelves were breaking apart. Measurements made directly on the ice showed that so-called streams, vast rivers of frozen water embedded in the WAIS, were speeding up to rates never seen before. The ice sheets appeared to be on the move.

By 2000 concern for the Greenland ice sheet was also growing. In 2006 satellite data showed a doubling of ice stream speed in southern Greenland. This concern was no idle speculation. Evidence existed that the end of the last ice age had seen the Greenland sheet breaking apart and spilling icebergs the size of cities into the ocean. Sea levels had risen rapidly immediately afterward. If something similar happened now it would be difficult if not impossible for modern society, with its major population centers lining the oceans, to deal with. "A ticking time bomb" is how one scientist described the situation.

The fate of the ice sheets in Antarctica and Greenland remains one of the most troubling uncertainties in climate science. But even if they prove more stable than the worst predictions we are in for difficult times ahead. After decades of intense scientific research, after decades of debate and conflicting views, a consensus is now in place. We are pushing hard on the dynamics of the Earth's climate, and that climate is a delicate system that could respond with frightening speed and force. Greenhouse warming and climate change are a reality. It is not a matter of "if" but "when" and "how much." Just seven years into the second millennium of the common era, human civilization finds itself facing, for the first time, a truly common threat of mythic dimension.

FROM MYTH TO SCIENCE AND BACK

FLOOD

Under the malicious glints of the clouds
the Kitakami, grown twice in width, perhaps ten times in
    volume,
bears yellow waves.
All the iron barges are being tugged to the army camp.
A motorboat sputters.
The water flowing back from downstream
has already turned into marshes
the paddies on the dried riverbed,
hidden the bean fields,
and devastated half the mulberries.
Gleaming like a snail's trail
it has made an island of the grass patch under the pines
and of the Chinese cabbage fields.
When and how they got there I don't know
but on the warm frightening beach
several dark figures stand, afloat.
One holds a fishnet.
I recognize Hosuke in leggings.
Has the water already
robbed us of our autumn food?
I climb the roof to look.
I hauled the manure bundles to a high place.
As for the plows and baskets
I went in the water a few minutes ago, up to my waist,
and managed to retrieve them.

        Miyazawa Kenji, *Miyazawa Kenji: Selections*

No one was around to watch the birth of the universe and compose a mythic narrative of its unfolding. Great cosmological myths have to spring from the imagination. People were, however, around for floods, lots of them. Whether the waters rose from increased sea levels at the end of an ice age or simply because a river swelled past its banks, countless generations of human beings have lived through flood and storm. The horror of watching children, parents, spouses, and siblings swept away must have been seared in personal, tribal, and cultural memories. Not all floods are great floods, of course. How long the memory of catastrophe persists depends on the extent of the devastation. Parents and grandparents still pass along stories of the great snow of '68 or the great flood of '49, but these stories fade with time and generations. Some experiences have a longer shelf life. The memory of the 1906 Galveston hurricane remains alive in songs that still have singers and audiences. The climate record indicates that rapid change has occurred in the lifetime of our species. That change may have brought such widespread devastation that its memory became more deeply sealed in legend, folklore, and myth.

Flood and storm are elemental human experiences. The transfer of those experiences into myth signifies their ability to touch the open-ended, mysterious quality of Being that I keep referring to as "the sacred." In his review of the many flood stories that exist, Alan Dundes of the University of California, Berkeley, makes a distinction between folktales and legends and myth. Myth, he says, is "a sacred narrative explaining how the world or humans came to be in their present form."[14] It is from that perspective that we can see how the emerging narratives of climate change function in the realm of the mythic.

In this book we have been pursuing a set of ideas that are inexorably intertwined. The first idea is that science and spiritual endeavor both emerge from an elemental experience of the world's sacred character. From that experience an aspiration, the constant fire, arises to understand and live in closer proximity to the world's fundamental power and mystery. The second idea is that myth represents our oldest response to that aspiration, and both science and religion retain living roots in mythology. Most important, science has a vital mythic function in modern culture. Human beings continue to have a basic need to interpret our experience of sacredness through narratives. These stories create meaning. They set our deepest longings and most deeply felt experiences in a sensible and spiritual context.

The narrative of climate change and global warming takes on the qualities of myth because it speaks to who we are as a race, what we have become, and where we are heading. By linking our fate with our actions and our relation to the great power that is the Earth, the climate change story also recovers the elemental forms of the flood myth. To see this let us begin where science always begins — with the data.

In the narrative of climate change the data were hard won. The record of temperature change came through real heroism on the part of researchers willing to endure the most inhospitable regions of the planet. The hero's journey for the sake of the community is manifest once again in the actions of men and women bundled in parkas struggling to make massive and massively complex equipment work high on the ice fields of Greenland and the Antarctic. From oxygen isotope ratios and trapped bubbles of paleoatmosphere, these scientists and their colleagues provided the raw material from which the narrative of climate science could be assembled.

Recall how our exploration of modern cosmology showed that science first seeks patterns in data and then creates narrowly focused narratives that explain those patterns. Climate science has its share of these more limited stories. These include how radiation is trapped by atmospheric CO_2; how salinity affects global circulation of the ocean; how changes in sea levels are driven by thermal expansion of the ocean. From the many threads of individual stories, a broader narrative is woven that addresses broader questions. What is the history of our planet's climate? Can human activity force dramatic changes in the Earth's climate system?

In seeing science's living roots in myth, what matters most is how the story gets told. It is the form and construction of the narrative that shows science serving our need for myth. To construct the grand narrative of climate change the narrow stories have been assembled into a kind of planetary epic that speaks of complexity and an almost infinite web of relationships between life and the thin veneer of air and water that supports it. Just as important is the story of our misdeeds, polluting a pristine world through thoughtlessness and greed and the inevitable catastrophic consequences. These are essential elements of myth and, in particular, flood myths. Like it or not, the narrative of climate change is one of sin, retribution, and great powers unleashed.

The public debate on global warming began in earnest in summer 1988. As early as June heat waves and drought hit the United States from the great cities of the East to the farmlands of the West. On June 23, a day when the Washington, D.C., temperature hit a record 93°F,[15] James Hansen, head of the Goddard Institute of Space Studies in New York, appeared before a congressional committee as an expert on climate change. He testified that based on his team's GCM simulations he was 99 percent sure that the march toward global warming was under way. He laid out the potential effects of climate change in no uncertain terms; the media listened even if Congress did not. Hansen's testimony was picked up by print and TV news outlets, and soon the debate over climate change had left the cloistered halls of academia and entered the public arena full force.

The issue immediately became polarized. The battle lines formed between environmental "greens" warning of damage to ecosystems and pro-development forces warning of damage to the economy. The science continued to progress throughout the 1990s, producing mounting warnings of the dire global consequences of inaction. The most strident opponents continued to claim that greenhouse warming was a hoax perpetrated by leftist elites and environmental radicals. The signing of the Kyoto Protocols by the majority of nations in 1998 helped close debate about the "truth" of global warming science. One hundred countries pledged to reduce CO_2 emissions in an effort to deal with the looming dangers of climate change. In spite of the absence of the United States from the Kyoto signatories, a popular consensus had emerged on the heels of the solidified scientific consensus. Humanity and its incessant fossil fuel consumption had messed with the planet in a way that might not be reversed.

By 2006 movies such as *The Day After Tomorrow*, with their dramatic special effects (in that case, Manhattan inundated by floodwaters), showed the extent to which the story had entered public consciousness. In spite of its silly vision of instant climate change *The Day After Tomorrow* made the moral message of global warming and human greed concrete. In his popular documentary on climate change, *An Inconvenient Truth*, former Vice President Al Gore repeats the phrase "This is a moral issue" a number of times. The mythic element in the narrative of climate change was now fully formed.

Gore's invocation of morality transcended his intention to inspire a sense of responsibility for the human suffering climate change is likely to incur. The emphasis on morality with respect to great deluges is a mythic theme. Retribution for the sins of humanity is a common element in flood myths. In a 1943 essay Hans Kelsen reviewed flood myths from across the globe.[16] Kelsen showed that, like the biblical story of Noah and the Babylonian story of King Atrahasis, a deluge is often set upon the people for their wickedness, laziness, or just plain annoying qualities. Many of the world's flood myths attribute great world-destroying storms to the gods' unhappiness with humanity's moral decay or its overpopulation and ensuing din. In the telling of our own impending flood myth, we have recaptured this theme. This occurs not simply because it may be that our activity increases CO_2 levels but also because the imprint of the myth of retribution via deluge remains with us.

Dundes makes the observation that attempts to understand flood myths generally fall into two categories: literal and symbolic. He describes the distinction: "Literalists tend to seek factual or historical basis for a given mythical narrative while advocates of one of the many symbolic approaches prefer to regard the myth as a code requiring some mode of decipherment."[17]

For Dundes the distinction is not mutually exclusive. "In the specific case of the flood there could, in theory, have been an actual deluge, either local or global, but at the same time one of the reasons why the flood narrative might have diffused as widely as it undoubtedly has — even to people far inland away from natural floods — could be attributable to its symbolic content." What matters most is that the myth provides a "sacred charter for belief." As Dundes says, "To the extent that the flood myth continues to be a vital force in [modern] life it remains a viable myth."[18]

In climate science the symbolic content of the myth is our confrontation with a sacred power — in this case the staggering complexity and interpenetration of the Earth's climate systems. Before 1988 environmentalism had already channeled this vision of the sacredness of the Earth, the sacredness of our place in and our responsibility for ecological systems. The emergence of the climate change threat came in the wake of recognition that we possessed power to dramatically change the environment. In the last half of the twentieth century, humanity remembered the living reality of its ancient mythic dreams of retribution and cataclysm.

The mythic vision of the Earth as a great coherent power became scientifically explicit in what became known as the Gaia hypothesis. Gaia was the Greek earth goddess, and hers was the name James Lovelock, a brilliant, iconoclastic scientist, gave to his maverick view of earth systems science. In the 1960s and 1970s Lovelock developed what was then a radical view that life did more than simply adapt to the nonliving environment. In the Gaia view, life changed the planet in significant ways — and to its own benefit. In 1974 he and the microbiologist Lynn Margolis published a paper titled, "Atmospheric Homeostasis by and for the Biosphere: The Gaia Hypothesis." The idea was simple: Over the course of time life on Earth had hijacked the planetary environment for its own ends. Under the Gaia hypothesis the atmosphere itself was a "contrivance" of life, its gases continually altered by biotic activity. In a sense the hypothesis held that the entire planet was a kind of living entity, like a giant self-regulating cell.[19]

The idea was initially ignored by most scientists. Many found the name enough to dismiss the Gaia hypothesis as mystical flimflam. In the decades that followed Lovelock and Margolis were careful in their explanations to show that life exerted no conscious intent in its maintenance of favorable planetary conditions. Eventually the use of *Gaia* was dropped for the more technical aphorism *earth systems*. At the same time the basic conclusion that the biosphere exerted tremendous influence on the planet's outer skin was recognized to be true. The issue of planetary self-regulation is still debated, but the vision of Earth as a complex interwoven system strongly driven by its living component survived. Most important, Gaia, as an idea, had been resurrected and filtered into popular consciousness. The story of climate science had inadvertently given a mythic name to its decidedly mythic narrative. Science grew from myth, and myth anticipated science.

Like scientific cosmology and the mythologies of cosmic genesis, climate science and flood myths represent paired, parallel responses to an elemental encounter with the world. The mythic response came first, embracing both observation of the world and emotive imagination. The scientific response has come only recently after the principles of scientific practice and proper mathematically based inquiry had been worked out. But to imagine the scientific response as devoid of the call of Mythos would be a mistake. In both scientific cosmology and climate science we have seen the retention of mythic themes. Most important, we have seen how these themes can animate the capacity of science to act as a hierophant, to make the sacred and the "awe"-ful character of the world stand forward and bring us face-to-face with an aspect of experience we all too often forget.

Science retains living roots in myth. Its particular power and capabilities may be a relatively new force in human history, but the source of that power and the origin of those capabilities have been with us a long time. Having explored the explicit relations between scientific and mythic narratives through these two examples, the questions that face us now are twofold. First, we ask if there is anything more we can say about the source of science's capacity for truth in light of its connection to the sacred character of experience. Then, finally, we turn to the question of the future. Does the connection between science, myth, and the sacred tell us anything we can use as we face into the bottleneck of our most dangerous century? These two questions are the subject of the last part of this book.

PART III

A New Path to the Waterfall

Science, Myth, Truth, and the Future

CHAPTER 8

Music of the Spheres

Truth, Myth, and Science

Genius, in truth, is little more than perceiving in an
unhabitual way.

William James, *Principles of Psychology*

If your knowledge of fire has been turned to certainty by
words alone, then seek to be cooked by the fire itself. Don't
abide in borrowed certainty. There is no real certainty until
you burn; if you wish for this, sit down in the fire.

Mevlana Rumi,
Daylight: A Daybook of Spiritual Guidance

I think I'll just
Let the mystery be.
Iris Dement,
"Let the Mystery Be"

The work had been interminable, exhausting, and fruitless. After
months of failed attempts to extract the chemical structure of benzene, the German chemist Friedrich August Kekulé was spent. He had
wasted another evening, and there was no good to come from it. There
was nothing else to do now but let it go. He turned his chair to the fire
and in the silence of his study drifted asleep. Then the dream came.

Kekulé's quest was the mid-nineteenth century's holy grail of chemistry.[1] Benzene is a hydrocarbon, a set of hydrogen and carbon atoms bonded together into a single molecule. It had first been isolated by the British physicist Michael Faraday in 1825 when he extracted a highly flammable gas from crude oil. Faraday discovered benzene when he distilled the slick film left over after the gas burned away.[2] Others soon found their own ways to isolate the substance. The discovery of benzene was part of a revolution in chemistry in the nineteenth century as a cornucopia of new hydrocarbons were discovered in rapid succession. Hydrocarbons became *the* hot topic among chemists. In the vast array of known chemicals, hydrocarbons were special because they appeared to form the sole molecular basis of life.[3]

Benzene was special among the hydrocarbons. It was prolific, forming the base of hundreds or even thousands of other chemicals while remaining remarkably stable. The sheer variety of benzene-like molecules overwhelmed chemists and pointed them to a facet of molecular behavior they could only guess at. Understanding benzene became the key needed to unlock the emerging science of organic chemistry.

The problem came down to structure, shape, and form. Faraday had already figured out what a benzene molecule was made of: six carbon atoms and six hydrogen atoms. What Faraday and scientists who followed literally could not see was how those atoms were arranged. The chemical promiscuity of benzene must have its source in the way atoms joined together. There were scientists during Kekulé's time who claimed it would be impossible to ever know the shape of molecules. Others were bolder, claiming that molecular structure could be known and used to drive progress in chemistry. Slowly a consensus grew. Finding the structure of benzene, without violating the known laws of chemical bonding, was moved high on the list of chemistry's priorities.

Kekulé never intended to study chemistry. He was born into a Bavarian noble family, who wanted him to study architecture. Murder led him in a different direction. In 1840, when he was a young architecture student, Kekulé took an apartment next to that of an old woman.

When she was found burned to death in her apartment her manservant was accused of the crime. Kekulé was called to the trial as a witness for the prosecution. The young man dutifully appeared in the courtroom and promptly met his own fate in the form of the great chemist Justus von Liebig. Liebig, a widely respected scientist and professor, had also been called to testify for the prosecution. The lawyers for the defense claimed the old woman had not been set on fire but had spontaneously exploded in flames after binge drinking. Using his expertise in chemistry, Liebig demonstrated to the court that it was impossible to consume enough liquor to saturate oneself into combustion. Leibeg's testimony not only helped convict the servant; it also made a convert of Friedrich Kekulé. He returned to the university, dropped architecture, and began chemistry studies in Liebig's lectures.

It was twenty years later, 1865, that Kekulé drifted into sleep by the fire. He had already become a professor at the University of Ghent in Belgium. For years he had worked on the structure of benzene, making small steps forward but no real progress. On that night, in that chair, everything changed. On that night Kekulé had a dream that moved him, and his science, forward in a single great leap. In the reverie of near-sleep Kekulé's quarry, the benzene molecule, appeared to him fully formed. As he described it in a lecture:

> I was sitting writing on my textbook, but the work did not progress; my thoughts were elsewhere. I turned my chair to the fire and dozed. Again the atoms were gamboling before my eyes. . . . My mental eye, rendered more acute by the repeated visions of the kind, could now distinguish larger structures of manifold conformation; long rows sometimes more closely fitted together all twining and twisting in snake-like motion. But look! What was that? One of the snakes had seized hold of its own tail, and the form whirled mockingly before my eyes. As if by a flash of lightning I awoke.[4]

Within weeks Kekulé confirmed the vision in his dream. Benzene was a configuration of six carbon atoms arranged in a hexagonal ring—a snake biting its own tail. The hydrogen atoms attached themselves at the vertices of the hexagon, creating a stick-figure snowflake. Structural chemistry had begun. Kekulé concluded his lecture with the now-famous admonition, "Let us learn to dream, gentlemen, and then we may perhaps find the truth."

. . .

Kekule's dream and his discovery of benzene's structure has been debated since 1890. Some have argued that he was not actually dreaming but in a semiconscious state of reverie. Others claim he was not the first to have the idea that benzene is a ring and therefore should not be considered its discoverer. Assuming he was not lying about the dream, it may be argued that Kekulé's dream was his psyche turning over data that his conscious mind had been struggling with for some time. What is not debatable is the nature and import of the dream image.

The snake creating a circle by biting its own tail is an ancient symbol called the Ouroborus that can be found in mythic imagery across the planet. In the popular interpretation of myth that Joseph Campbell espoused (borrowing from Jung), the Ouroborus is an archetype. It is a fundamental form, repeated in the graphic depiction of mythic narratives over and again in cultures as diverse as the ancient Egyptians, the Mesoamerican Incas, the expansive Chinese dynasties of the first millennium, and medieval Europeans. As an archetype, the Ouroborus is claimed to be a symbol that carries a meaning, a truth, that extends beyond individual representation. Truth, transcending time and place, transcending history, is the core idea behind archetypes. The fascination of Kekulé's dream lies in the link between the archetypal image of the Ouroborus and a specific arrangement of carbon atoms in a specific molecule.

Kekulé's dream is iconic for us because it touches the deepest issues our exploration of myth, science, and the sacred can raise. Is there any truth in myth, via archetypes or other forms, that rises above history and speaks directly to something stable and enduring? If so, how does science, with its living roots in myth, relate to these truths? This is one of the big questions striking at the heart of what makes science so effective and the nature of the truth it uncovers. To define the question we will have to break it into parts, starting with the nature of truth in science.

SCIENCE, TRUTH, AND MATHEMATICS

The allure of science as a path toward understanding, toward truth, is its ability to rise above the specifics of time and place. The laws of nature, discovered through the scientific process, stand outside of history. The chemical pathways of photosynthesis, for example, nourish the houseplants on your windowsill today just as they did on the alluvial plains of Jurassic swamplands 140 million years ago. Newton's law of gravitation was true this morning on the corner of Thirty-fourth Street and Broadway in New York City, and it was true three hundred years ago at the Tiananmen Gate of Ming dynasty China. More to the point, the law of gravity operates "right now" in galaxies spread like pollen spores across space and time.

The laws of nature appear to be just that, laws that are absolute and timeless. By siting a spiritual complement of science in myth we could hope to find a complementary stability in its truths that might also rise above the specifics of history. Does myth reveal any transhistorical truth in a way that sheds light on its relationship with science? To reach across such a broad question we must first become familiar with the deepest source of enduring truth in the physical sciences — the compelling and beautiful domain of mathematics.

There is a long-standing debate among scientists about the nature of mathematics and its ability to represent the world. For many researchers mathematics is a kind of scientific secret language built into nature. In this view the mathematics that animates science points us to an invisible reality that underlies everything we experience. For other scientists mathematics is something we simply invent. In their eyes mathematics says more about the evolution of the human brain and its ability to respond to patterns than any objective, independent reality. There is no debate, however, about the unique and uniquely powerful role of mathematics in science. More than anything else, mathematics gives science the appearance of achieving eternal, universal truth. In this sense it provides an archetype of archetypes, an example of truths that stand outside of history.

In 1960 Eugene Wigner, a physicist well known for his contributions to quantum theory, published an essay titled "The Unreasonable Effectiveness of Mathematics in the Natural Sciences."[5] He begins the essay with a quote from Bertrand Russell:

> Mathematics, rightly viewed, possesses not only truth, but supreme beauty, a beauty cold and austere, like that of sculpture, without appeal to any part of our weaker nature, without the gorgeous trappings of painting or music, yet sublimely pure, and capable of a stern perfection such as only the greatest art can show. The true spirit of delight, the exaltation, the sense of being more than Man, which is the touchstone of the highest excellence, is to be found in mathematics as surely as in poetry.

This call to see mathematics as a source of sublime experience frames Wigner's own sense of the subject. A good part of his essay articulates his astonishment that mathematics, a language seemingly invented by humans, should appear as a fundamental structure of reality in physics. Wigner begins with the observation that "mathematical concepts turn up in entirely unexpected connections. Moreover, they often permit an unexpectedly close and accurate description of the phenomena in these connections." The connections between mathematics and the phenomena of physics are by no means obvious from the outset. As Wigner describes it, "We are in a position similar to that of a man who was provided with a bunch of keys and who, having to open several doors in succession, always hit on the right key on the first or second trial." For Wigner this is more than surprising. He marvels at the centrality of mathematics in the natural world: "The enormous usefulness of mathematics in the natural sciences is something bordering on the mysterious and there is no rational explanation for it."

Wigner then tries to define his subject: "Mathematics is the science of skillful operations with concepts and rules invented just for this purpose. The principal emphasis is on the invention of concepts." He turns his scrutiny on mathematics as an invention of the mind: "[While] it is unquestionably true that the concepts of elementary mathematics and particularly elementary geometry were formulated to describe entities which are directly suggested by the actual world, the same does not seem to be true of the more advanced concepts, in particular the concepts which play such an important role in physics." Mathematics on its own is, according to Wigner, an exercise in abstract thinking with no inherent connection to the world. He cannot imagine it is something we evolved into. "The great mathematician fully, almost ruthlessly, exploits the domain of permissible reasoning and skirts the impermissible. . . . [C]ertainly it is hard to believe that our reasoning power was brought, by Darwin's process of natural selection, to the perfection which it seems to possess."

Wigner then considers how often the most arcane pieces of pure mathematics find their way into the heart of physics. "It is difficult to avoid the impression that a miracle confronts us here," he claims. To illustrate his point he begins with the simple example of Isaac Newton and his predecessor, the enigmatic Johannes Kepler (1571 — 1630).

Kepler was an astrologer, an astronomer, a mathematician, and a dedicated Copernican. Kepler's great contribution to astronomy was to establish the true laws of planetary orbits, including their shape. For millennia astronomers were sure that all planetary motion amounted to the traversal of perfect circles. Even when Copernicus overthrew the geocentric solar system he kept his planets moving in perfect circles with the Sun as their center. The dedication to circles as the geometry of the heavens was a philosophical bias handed down from the Greeks. There was no way, however, to reconcile circular orbits with what was actually seen in the sky, so even Copernicus's model produced errors when predicting where any given planet would be on any given night. Kepler set out to find the true form of planetary motion. After years of work he found that all planets traveled along *elliptical*, not circular, orbits. The ellipse is really a family of geometrical figures known to the Greeks from their explorations of pure mathematics. The shape of individual ellipses can range from circles to extremely thin, cigar-like shapes. The key point is that *all these shapes can be described with a single equation*. The circular ellipse and the cigar-shaped ellipse and everything in between are hidden in the mathematical formula for "the ellipse." If one is so inclined, the poetic economy of this one equation serves as an example of mathematical beauty.

Using this one equation, Kepler found, he could embrace the motion of all the known planets. All the data, all the carefully charted positions of Mars, Jupiter, and the rest, were neatly and accurately accounted for with Kepler's simple, single equation for elliptical motion. For Kepler it was an almost mystical example of the harmony of mathematics manifesting itself in the world. One hundred years later Newton drove that harmony to a deeper level, showing that all gravitational forces would naturally lead to elliptical motion. This is the kind of unreasonable effectiveness of mathematics that Wigner is taken by. "It was Newton," Wigner writes, "who brought the law of freely falling objects into relation with the motion of the moon, noting that . . . the thrown rock's path on the earth and the circle of the moon's path in the sky are particular cases of the same mathematical object of an ellipse, and postulated the universal law of gravitation on the basis of a single, and at that time very approximate, numerical coincidence." The ellipse, an abstract mathematical object, turns out to be the single form of motion applicable to everything from the falling apple to the far-flung comet. The applicability of ideas conjured up by mathematicians for their own ends is the essential question for Wigner. "Fundamentally," he says, "we do not know why our theories work so well."

At the end of the essay Wigner sums up his perspective:

The miracle of the appropriateness of the language of mathematics for the formulation of the laws of physics is a wonderful gift which we neither understand nor deserve. We should be grateful for it and hope that it will remain valid in future research and that it will extend, for better or for worse, to our pleasure, even though perhaps also to our bafflement, to wide branches of learning.

Wigner claims we do not understand why mathematics is so effective in science; others are not so agnostic. The foundations of modern mathematics were set down about 2,500 years ago in classical Greece. At the same time Plato, the greatest philosopher of that age, provided an account for mathematics' mysterious efficacy. Plato believed that mathematics was the model for understanding all reality. Above the entrance to the school he founded in Athens he inscribed the famous motto, "Let No One Ignorant of Geometry Enter." For Plato the truths of mathematics did not relate to actual physical objects such as squares, circles, and triangles. Instead they referred to idealized entities that existed in another reality, the so-called Platonic realm of idealized forms. The world of the idealized forms was separate from ours, existing "below" or "beyond" it. As the highly respected mathematical physicist Roger Penrose puts it, "Physical structures such as squares, circles and triangles cut from papyrus or marked on a flat surface . . . might conform to these ideals very closely but only approximately. The actual mathematical squares, cubes, circles, spheres, triangles, etc., would not be part of the physical world but would be inhabitants of the idealized world of forms."[6] Thus our world is just a shoddy shadow of the Platonic realm. This second-rate status is the price we pay for understanding why mathematics is so effective in science. Our world is built off these ideal forms. It is the flesh hung on the ideal skeleton. Thus all mathematical truth serves as a preexisting perfect prototype, an archetype, of all we experience.

The reality of the Platonic realm has been debated for two thousand years. After all that time you might think the issue has grown crusty or that someone would have worked out the "right" answer. It is a testimony to Plato's insight and the murkiness of metaphysical questions surrounding mathematics' effectiveness that the issue remains so potent.

There is an alternative explanation for Wigner's unreasonable effectiveness of mathematics, one that does not postulate a separate, ideal realm of existence. For many scientists the mathematics we have is the only one our brain and its evolution allows us to have. One modern example of this approach can found in *Where Mathematics Comes From* by George Lakoff and Rafael Nuñez.[7] Their perspective derives from modern cognitive science. It emphasizes that all mathematical concepts emerge from our experience as "embodied minds." We are animals whose brains evolved to deal with the one real, physical world. Our minds only exist as part of our bodies. Mathematics is a kind of metaphorical thinking, derived from the experience of living in our bodies. Thus experience of the physical world is the one and only source of mathematics and mathematical truth. According to Lakoff and Nuñez, "Since human mathematics is all we have access to, mathematical entities can be only conceptual in nature arising from our embodied conceptual systems."[8] Since mathematics emerges from our own physicality, it is no surprise that mathematics is not built into the universe. For Lakoff and Nuñez and others who take such a position there is no Platonic mystery in mathematics and no alternative, transcendent reality inhabited by mathematical forms.

The enduring truth of mathematics, and hence science, can be seen either as a kind of transcendent archetype or as something encoded in the structure of our brains. This dichotomy might not seem unfamiliar. We have already encountered it in another set of clothing. The theories of myth discussed in chapter 5 showed us a similar discussion among comparative mythologists.

ARCHETYPES, STRUCTURALISM, AND CRITICS OF THE UNIVERSAL

More than one of the great thinkers in comparative mythology has suggested that myth contains some form of transhistorical truth about human Being and the world. These ideas are as contentious as they can be compelling. They must be treated carefully and with a critical eye. Still, given the connections between myth and science already laid out, the siren call of universals is too strong to ignore just because they are likely to offend some sensibilities. With one eye on their pitfalls and the other on their promise, let us begin our own exploration of enduring truths in the interwoven realms of science and myth.

When Jung first proposed the concept of archetypes in myth and dreams he saw them as existing in what he called a "collective unconscious," a "pan-human pre-cultural stratum." Archetypes existed in this stratum and were expressed by individuals and cultures in the creation and maintenance of myths. For Jung the human mind is not born as a blank slate but carries the imprint of forms in the collective unconscious as a kind of instinct. In Jung's words, "The psychological manifestations of the instincts I have called archetypes . . . are by no means useless archaic survivals or relics. They are living entities which cause the formation of numinous ideas."[9]

Numinous is, in a sense, another word for the sacred. For Jung the great themes and symbols in myth are repeated so often and in so many cultures because they are explicit archetypes. They speak to the deepest human experiences of life and transcend any individual or culture.

The archetypes and the collective unconscious have proven to be a litmus test for those who study myth. Depending on one's inclination they can provide scholarly window dressing for notions of a nonphysical plane of existence. As Alan Dundes puts it, "There is an unquestionably mystical, anti-intellectual aspect of Jung's thought. Since the archetypes are part of the collective unconsciousness they cannot . . . ever be made fully conscious. They are not therefore completely susceptible to rational definition or analysis."[10] Dundes then quotes Jung directly to make his point: "Contents of an archetypical character are manifestations of processes in the collective unconscious. Hence they do not refer to anything that is or has been conscious but to something which is essentially unconscious. In the last analysis it is impossible to say what they refer to." Jung would defend his understanding of the archetypes by maintaining their relationship to something less controversial — human instincts. "It is important to bear in mind," Jung writes, "that my idea of the archetypes has frequently been misunderstood as denoting inherited patterns of thought or as a kind of philosophical speculation. In reality they belong in the realm of the instincts and in that sense they represent inherited forms of psychic behavior."[11]

Joseph Campbell drew heavily on the archetypes in his own thinking about myth. For Campbell archetypes were indeed universal forms: "The archetypes, or elementary ideas, are not limited in their distributions by cultural or even linguistic boundaries, they cannot be defined as culturally determined."[12] According to him, the archetypes exist in service to myth's distinctly mystical function. All myths told the same story and taught the same lesson. In his analysis of Campbell's work, Robert Segal explains that in Campbell's view myths proclaim "the oneness of all things. Myths not only assume, but even preach mysticism. Myths proclaim that humans are one with one another, with their individual selves and with the cosmos itself."[13] This is one of the principal facets of Campbell's work that has given it such general appeal.

The unity of life is both high idea and grand ideal. As Segal says, "No tenet is more staunchly romantic than the conviction that beneath the apparent disparateness of all things lies unity."[14] In Campbell's work the archetypes are stable forms within myth that communicate this unity. Campbell has been criticized because he claimed myth's only function was to show the world's unity but never proved how this claim could be true. For rationalists, including many scientists, any strong claim of mystical unity is enough to set eyes rolling and bring the argument to a crashing close. It is worth noting, though, that physicists feel the same pull toward unity in their attempt to tie the four forces in a single grand unified force. In that case the issue can be laid out with the discursive tools of rational investigation, of course. Still, one can certainly ask why scientsits pursue the issue with such passion and conviction in the first place.

Another criticism of archetypes (via Jung or Campbell) is that seeing myth in terms of universals abstracts them from the history of their actual creation. Archetypes seem so potent because they offer something that transcends time, place, and the specifics of culture much like the laws of science. Modern critics of the idea of archetypes respond that unlike Newton's laws or the chemistry of benzene, archetypes do nothing more than solidify the cultural prejudices of the myth's interpreter, namely, modern Westerners. In her insightful work on comparative mythology, *The Implied Spider*, Wendy Doniger quotes one of these critics, Maria Werner, who explores the role of gender and archetype: "When history falls away from a subject we are left with Otherness and all its power to compact enmity. . . . An archetype is a hollow thing but a dangerous one."[15] Werner looks askance at writers who claim, for example, that the evil stepmother is an archetype (Cinderella, etc). By calling this construct an archetype, the historical reasons for women's cruelty within the home are erased. Modern scholarship, with its emphasis on power and politics, has rejected the archetype along with Campbell's grand idea of a monomyth as intellectual colonialism that levels the specifics of individual cultures and their stories in the name of a grand abstraction.

These criticisms have validity. While Jung remains a powerful and progressive force in psychological thinking, I have met some followers of Jung and Campbell who could easily find themselves nodding in agreement with the most wide-eyed New Age enthusiasts. All critical sensibilities have been thrown to the wind. Postulating the existence of a metaphysical entity called the collective unconscious that can be described enough to pull out archetypes but not enough to specify where or how they exist is a bit like trying to have your cake and eat it too. Political dimensions inherent in the grand ideals of universals must also be kept in mind because we must always be honest about our own biases. As Doniger points out, "What we regard as archetypical, universal or even natural ('the way things are') is not immutable or desirable; it is merely given."[16]

Both criticisms must be kept in mind, but they do not require us to abandon the notion of archetypes in our exploration of science and the sacred but to use it cautiously. There is precedent for this: even Eliade used the concept of archetype for a time in his descriptions of myth, ritual, and the sacred. "Every ritual has a divine model, an archetype," he claimed.[17]

> The very dialectic of the sacred tends to repeat a series of arche-
> types, so that a hierophany realized at a certain historical moment
> is structurally equivalent to a hierophany a thousand years earlier
> or later. Hierophanies have the particularity of seeking to reveal
> the sacred in its totality, even if human beings in whose conscious-
> ness it "shows itself" fasten upon only one aspect or small part of
> it. In the most elementary hierophany everything is declared.
> The manifestation of the sacred in a stone or a tree is neither
> less mysterious nor less noble than its manifestation as a "god."
> The process of sacralizing reality is the same: the forms taken
> by the process in man's religious consciousness differ.[18]

Myths were always a sacred retelling of acts associated with origins that occurred in a distant timeless time. All human activities had mythic prototypes and these were, for Eliade, a kind of archetype. Eliade uses the examples of the arts, dance in particular, to illustrate the point: "All dances were originally sacred, in other words they all had an extra-human model. . . . Choreographic rhythms have their model outside of the profane life of man whether they reproduce the movements of the . . . emblematic animal or the motions of the stars . . . — a dance always imitates an archetypical gesture or commemorates a mythical moment."[19] Eliade did, however, want to distinguish his use of arche-type from Jung's. Doniger points out that in the preface to the English edition of his *Myth of Eternal Return* Eliade states, "I nowhere touch upon the problems of depth psychology nor do I use the concept of collective unconscious. I use the term archetype as a synonym for exemplary model or paradigm."[20]

The archetypes of Jung and Campbell are not the only expression of myth embracing deeper transcendent truths. In exploring theories of myth we encountered another form of universalism in the work of Claude Lévi-Strauss. Lévi-Strauss's structuralism offered a different way for myth to rise above the specifics of history and expose fundamental aspects of human being. What transcended any individual myth for him was not a set of archetypal symbols existing in a collective unconscious but a narrative structure that revealed the architecture of the human mind. That architecture was, for Lévi-Strauss, built around the conflict of opposites. In his view the brain tried to take all the disparate data fed to it and make sense of them in binary terms. It was as if human consciousness could only structure the world in terms of pairs of opposing ideas: night and day, birth and death, the high places of gods and the low places of demons. This structuralism is more attractive to the rationalist mind-set. It does not suggest transpersonal, nonphysical connections between individual psyches the way Jung's account might be interpreted. Doniger provides a concise description of what structuralism tells us about myths and how to look at them: "Structuralism gives us a pretty good idea of how myths are made. If you take an old story . . . .and compare it with a later telling it is as if the first story were dropped and broken into pieces, then put back together again differently—not wrongly, just differently. The pieces are the atomic units of myth, what Levi-Strauss called the mythemes."[21]

In the structuralist view every myth will use some of these atomic units. What matters for Lévi-Strauss is how the human brain, structured to deal with the logic of opposites, takes these mythemes and uses them to express an internal binarity that it cannot escape. For Lévi-Strauss myth is a form of language, and "language predisposes us to understand ourselves and our world by superimposing dialectics, dichotomies, or dualistic grids on data which may not be binary at all." The presence of these "mythemes" expresses the basic structure of the brain (for Lévi-Strauss) via the formula of binary opposites and their reconciliation. In structuralism this formula is what allows a myth to remain the "same" even when it undergoes the major transformations of different cultures and different times. Using this understanding as a base, Lévi-Strauss hoped that the logical themes of myth could be described with an almost mathematical precision. Doniger quotes a passage from Lévi-Strauss that reveals this quasi-scientific emphasis: "I am convinced that the number of these [mythic] systems is not unlimited and that human beings . . . never create absolutely; all they can do is choose certain combinations from a repertory of ideas which it should be possible to reconstitute. For this one must make an inventory of all the customs which have been observed by oneself and others."[22]

By "customs" Lévi-Strauss also means stories and myth. Here he makes a point similar to the one Marcelo Gleiser raises about cosmology. There are only so many ways to tell our fundamental stories, and myth contains them all. Lévi-Strauss's inventory project is an ambitious and probably impossible project, however. Like the grand vision of archetypes, structuralism is ultimately a victim of its own grand designs.

As Doniger points out, structuralism, like the archetypes, has been criticized for ignoring history. Like all grand theories it plows over what actually happened at one time and place to level the field for a transcendent nonhistorical "always happening everywhere." In addition, structuralism has been criticized for bleeding all meaning and life from myth. Doniger quotes one Lévi-Strauss critic: "Aloof, closed, cold, airless, cerebral . . . [Lévi-Strauss's] books seem to exist behind glass. Self-sealing discourses in which jaguars, semen and rotting meat are admitted to become oppositions, inversions, and isomorphism."[23] In other words the embodied reality of people making myths is stripped down to a skeleton of abstractions.

But Doniger is not ready to give up on structuralism, just as she is not ready to give up on the project and promise of comparative mythology. There are important things that can be learned in looking for similarities. Intention and method are what matter most for her. Her emphasis on *intention*, a word close to *aspiration*, provides an instructive way of looking at both archetype and structuralism in our discussion of myth and science. For Doniger there are no monomyths, but there are micromyths (local variations) and macromyths (larger mythic themes). She likens structuralism to a bus: it can take you quite far in understanding myth if you know when to exit. "We must jump off Lévi-Strauss's bus one stop before he does," she writes. "In order to remain fully engaged with our texts [specific myths] we must wallow in the mess for a while before the structuralists clean it up for us."[24]

We can now look at archetypes, structuralism, science, and myth on the larger scales of universal truths. Sound bites will not do in this territory. We remind ourselves, as the Zen Buddhists do, that our job is to keep from fooling ourselves with easy answers. In this spirit we begin by revisiting history's most extraordinary encounter between myth, science, archetype, and physical law — the thirty-year correspondence between Carl Jung and Wolfgang Pauli.

ARCHETYPES, ATOMIC PHYSICS, AND WOLFGANG PAULI

I was still wet behind the ears and did not yet know better. My honors project professor was a man of few words and very intimidating. For three months I had been working under his direction. My project was to use a simplified form of general relativity to describe tiny deviations in the Moon's orbit around the Earth. General relativity is notorious for its huge, laborious calculations. Producing even relatively simple results requires pages and pages of algebra. Making sure that the product of the 12th term on page 2 and the 8th term on page 5 were correct and correctly copied as the 34th term on page 19 was an exercise in mindfulness worthy of a monk. My monkhood wasn't going so well.

I kept getting the wrong answer. Well, what I mean is, I kept getting an answer that differed from what we had expected. There were other ways to get the result, and the point of the exercise was to show that this form of general relativity could reproduce an answer others had found using traditional approaches. My undergraduate project was supposed to be a check on the new equations, but my result differed from the correct answer by a factor of 4/37 and a term with a cosine in it. Ugh! After checking and rechecking, I had a bright thought. "Perhaps this was a new result," I told myself. "Maybe I had found something no one else knew about!" With my notebook of calculations in hand and a stomach full of trepidation I went to see my professor.

In the quiet of his office and surrounded by a heavy pall of pipe tobacco smoke, he looked over my work for what seemed like a very long time. Then he looked up and smiled.

"Have you ever heard of Wolfgang Pauli?" he asked. I told him I had heard of the Pauli exclusion principle in quantum mechanics. "Good," he replied. "Pauli was a great and uncompromising theoretical physicist. He was called the conscience of physics for good reason. Pauli was the critic everyone counted on to see into the heart of a calculation. Once he was given a paper by a colleague to review. After looking it over Pauli said, 'This paper is so bad it is not even wrong.' It's a good lesson for you to learn because the same can be said about your calculation." My professor then, kindly, went on to show how a mistake I had made early on had led the derivation astray. Eventually I got the calculation right, but I never forgot my professor's reverence for Pauli or the incisiveness of Pauli's barb. "Not even wrong." Ouch.

Wolfgang Pauli is a hero to many scientists, including me. He was a prolific physicist who was honored for his work at the highest levels, including the Nobel Prize in 1945 for his exclusion principle, a law of subatomic physics used in everything from electronics to the study of neutron stars.[25] For those of us who only know Pauli as the paragon of theoretical physics it is nothing less than stunning to find he is the author of the following letters.[26]

Zurich, July 7, 1937

Dear Professor Jung

I would just like to say a brief word of thanks for sending me your treatise on alchemy. It was bound to be of great interest to me, both as a scientist and also in light of my dream experiences. These have shown me that even the most modern physics also lends itself to the symbolic representation of psychic process, even down to the last detail. Of course nothing is further from the thoughts of modern man than the idea of penetrating the secrets of matter this way.

Alchemy! What was the conscience of physics doing mucking about with alchemy? In another, more surprising letter Pauli writes to Jung about the motivation for his inquiries.

Dear Professor Jung

After a careful and critical appraisal of many experiences and arguments, I have come to accept the existence of deeper spiritual layers that cannot be adequately defined by the traditional concept of time.

This language and these topics are not part of the canonical image of Pauli. He was, surprisingly, a man of surprisingly diverse sensibilities. He was in Jungian therapy for only a few years after the crisis of his failed marriage (see chapter 4), but he and Carl Jung remained in contact for most of the rest of their lives. Pauli's dreams, which were cinematic in their imagery, led him to a lifelong exploration of archetypes and their appearance in physics. Pauli literally dreamed in physics. The language of these dreams convinced him that concepts of theoretical physics were themselves a kind of archetype. To Pauli, physics tapped deep resonances with the totality of human experience, including the experience of the sacred.

Pauli's dreams served as the starting point for his exploration of what he called "background physics." Pauli coined this term to describe the appearance of "physical terms as archetypal symbols." In an unpublished essay he described this background physics, giving us a clear view of his experience and its influence on his thinking:

Under background physics I understand the occurrence of quantitative notions and conceptions of physics as spontaneous fantasies [i.e., dreams] in a qualitative and figurative — i.e., symbolic — sense. The existence of this phenomenon has been known to me for 12 or 13 years from my own personal dreams, which are totally

uninfluenced by other people. . . . As befits my rational scientific
approach these dreams seemed to me initially offensive — in fact
an abuse of scientific terminology. . . . Later however I came to
recognize the objective nature of these dreams or fantasies — i.e.,
the fact they were largely independent of the actual person. What
first struck me was the similarity of the mood that obtains in my
dreams and in the physical treatises of the 17th century, especially
in Kepler, where scientific terms and concepts were still relatively
undeveloped and physical consideration and ideas were interspersed
with symbolic concepts.[27]

A dream from 1946, described to Jung in a letter, illustrates the
fecundity of Pauli's unconscious imagination and the way physical concepts manifested themselves. A recurring figure in Pauli's dreams was
"the Blond," a tall man somewhat younger than Pauli. Pauli's description of the dream begins:

> The Blond stands next to me. In an ancient book I am reading
> about the Inquisition's trials against the disciples of the teaching
> of Copernicus (Galileo, Giordano Bruno) as well as about Kepler's
> image of the trinity. Then the Blond says "the men whose wives
> have objectified rotation are being tried." This upsets me greatly.

Later in the dream Pauli finds himself on trial. Then the Blond is next
to him again as he reads the ancient book.

> Then the Blond says sadly (apparently referring to the book),
> "The judges do not know what rotation or revolution is, and that
> is why they cannot understand the men." With the insistent voice
> of a teacher he goes on to say "But *you* know what rotation is!"
> "Of course!" comes my immediate reply., "The circulation of the
> blood and the circulation of the light — all that is part of the basic
> rudiments." . . . Whereupon the Blond says "Now you understand
> the men whose wives have objectified their rotation for them."[28]

This was a guy who knew how to dream. It is hard not to be
impressed by the sheer gravity of Pauli's imagination. Pauli sensed that
his dreams extended beyond his personal story or his own conception

of physics. The symbols he dreamed in — frequency, dipoles, spectral line splittings — were elevated beyond their usual textbook, research meanings. In time Pauli came to realize that these dreams and symbols were a gateway to the exploration of his sense of the sacred as it appeared through physics. The clarity of Pauli's recognition of the explicit nature of scientific ideas as hierophanies is remarkable considering the immense role he played in the development of modern physics.

Describing his response to the dream of the Blond and the book, Pauli writes, "There upon I woke up very shaken. The dream was an experience of a numinous [unspeakable, mysterious, frightening] character which influenced my conscious attitudes in an essential way."[29] It was in response to these dreams that Pauli took up the study of archetypes in the development of physical science. His focus in this work fell on Johannes Kepler.

The scientific, the symbolic, and the sacred were intertwined in the work of Kepler, a kind of scientific mystic. For him mathematics embodied an eternal order made manifest in the world. His discovery of elliptical planetary orbits made after years of study filled him with a sense of rapture. This made him an attractive figure to Pauli in his attempt to weave together the themes of science, symbols, and the sacred.

In the late 1940s and early 1950s Pauli began a study of Kepler that culminated in an essay titled "On the Influence of Archetypal Ideas on the Scientific Theories of Kepler." Pauli also presented his thoughts on Kepler in lectures to the Psychological Club of Zurich under the same title. He explained his interest in Kepler as follows:

> My attention was therefore directed especially at the 17th century when, as the fruit of great intellectual effort, a truly scientific way of thinking, quite new at the time grew out of the nourishing soil of a magical-animistic conception of nature. For the purpose of illustrating the relationship between archetypical ideas and scientific theories of nature, Johannes Kepler seemed to me especially

suitable since his ideas represent a remarkable intermediary stage between earlier magical-symbolical and modern quantitative mathematical descriptions of nature.[30]

Pauli could easily have substituted "mythic thinking" for "a magical-animistic conception of nature" in this passage since his emphasis was on symbolic archetypes with mythic roots. In the essay Pauli goes on to explore relationships between "intuition and the direction of attention" as it pushed forward the development of key concepts in the physical sciences. The "direction of attention" for Pauli meant the specific way in which symbolic, archetypal images acted as guides for creative scientific theorizing. This was a radical idea. To forge links between the raw material of the senses and the development of concepts in physics, Pauli was willing to go out on a limb (at least for a physicist). He states:

> It seems most satisfactory to introduce at this point the postulate of a *cosmic order independent of our choice and distinct from the world of phenomena.* This order manifests itself in a matching of inner images pre-existent in the human psyche with external objects and their behavior. . . . These primary images . . . are called by Kepler archetypes. . . . Their agreement with the primordial images or archetypes introduced into modern psychology by CG Jung and functioning as instincts of the imagination is very extensive.[31]

In his exposition of Kepler's works Pauli finds numerous symbolic archetypes. These were the inspiration that forged Kepler's intellectual will to pursue Copernican models. Pauli writes:

> The symbolical images and archetypal conceptions are what cause [Kepler] to seek natural law . . . . Because Kepler looks at the Sun and the planets with this archetypical image in the background, he believes with religious fervor in the heliocentric system — by no means the other way around as a rationalistic view might cause one to erroneously to assume.[32]

In other words Kepler's connection with mythic thinking, developed through exploration of symbolic archetypes, drove him to his specific scientific convictions. He did not compare solar system models first and then dispassionately decide among them based on the scientific method. Pauli concludes his study of Kepler with a description of the current state of affairs:

> It is obviously out of the question for modern man to revert to the archaic point of view that paid the price of its unity and completeness by a naive ignorance of nature. [Modern man's] strong desire for a greater unification of his world view, however, impels him to recognize the significance of the pre-scientific stage of knowledge for the development of scientific ideas . . . by supplementing the investigation [of physics] with this knowledge directed inward.[33]

Pauli clearly believed that a sacred, experiential dimension of reality existed, and he also believed he had found a means to understand at least some aspects of its character.

Reviewing Pauli's work in the arena of archetypes and physics is like standing at the edge of a cliff. There have certainly been Nobel Prize winners who have gone off the deep end, spending their remaining years focused on fruitless explorations of parapsychology and ESP. What sets Pauli apart is the depth of his scholarship, the rigor of his sensibilities, and the range of his sensitivities. Pauli never stopped his work at the frontiers of physics. He continued to make significant contributions to particle physics until his death. His exploration of archetypes in physics, via Kepler, shows the same voracious appetite to know the workings of the world in detail.

Wolfgang Pauli seems unique among modern scientists in his profound understanding of the machinery of science and the role it plays as explicit hierophany. This understanding led him to a deeper experience of the world's sacred character. I am not prepared to follow Pauli in his claim that scientific concepts exist in a Jungian collective unconscious. Pauli does, however, point a way for us to think more broadly about science and the sacred. His experiences and his deeply thoughtful response to them raise the standard of the discussion. What then can we find in the interwoven strands of science and myth that can illuminate the great and constant aspiration to know the truth?

A UNIVERSE OF UNIVERSAL STORIES

I had just come from my undergraduate partial differential equations class and was in serious need of caffeine. We had completed our fourth straight day of lectures on the equations of a vibrating membrane. My head hurt, and my hands were cramped from taking notes. Partial differential equations (PDEs) appear everywhere in mathematical physics. They provide scientists with the language to describe the evolution of collapsing clouds of interstellar gas, the nature of oscillating electromagnetic fields, and the flow of traffic on a four-lane highway. By solving these equations in all their abstract glory the behavior of the real system can be predicted, described, and *understood*. It was very cool.

The going was tough, though. Like constructing an invisible house of cards, we had spent the last few days building up a story based on theorems and postulates. Then, finally, we had enough background to really get started. The vibrating membrane was a general problem. The membrane could be a drumhead, the surface of a lake, or the surface of a star. The professor taught us to use simple vibration patterns as a kind of grammar. He showed us how to add up these simple patterns and describe complex oscillations. Imagine, for example, the quick smack of a drumstick on a drum. Using what we had learned we could, exactly and explicitly, describe every detail of the drumhead's complex, evolving pattern of vibration by adding up lots of simple patterns.

I had filled half a notebook during the lectures. Now I was tired and needed caffeine. In the student cafeteria I got a Styrofoam cup, filled it, and got in line to pay. As I searched for my wallet I put the cup down on an ice-cream freezer. After extracting the needed $1.25, I reached for the cup and was stopped dead in my tracks. There it was, laid out with exquisite perfection, right in front of me.

The freezer was gently vibrating, set in motion by its small motor.

As it rested on the freezer, my coffee cup picked up the oscillations. On the coffee's surface I saw the exact pattern I had just learned about in class. The ordered flow of the surface reflected fluorescent light from above, revealing tiny circular ripples superimposed on crisscrossed radial stripes. The pattern was complex but ordered and stable. Ten minutes ago I had seen the same pattern represented as a long string of mathematical symbols or as a diagram on graph paper. Now it was real. Now it was "true." Suddenly the abstractions were alive for me. The mathematics was made manifest in motion. It was one of the most beautiful things I had seen or ever would see. There was a long moment before I was willing to exhale and get on my way.

• • •

CHOOSE SOMETHING LIKE A STAR
O Star (the fairest one in sight),
We grant your loftiness the right
To some obscurity of cloud —
It will not do to say of night,
Since dark is what brings out your light.
Some mystery becomes the proud.
But to be wholly taciturn
In your reserve is not allowed.
Say something to us we can learn
By heart and when alone repeat.
Say something! And it says, "I burn."
But say with what degree of heat.
Talk Fahrenheit, talk Centigrade.
Use language we can comprehend.
Tell us what elements you blend.
It gives us strangely little aid,
But does tell something in the end.
And steadfast as Keats' Eremite,
Not even stooping from its sphere,
It asks a little of us here.
It asks of us a certain height,
So when at times the mob is swayed
To carry praise or blame too far,
We may choose something like a star
To stay our minds on and be staid.
                                        Robert Frost

The constant fire is the aspiration to know what is essential, what is real, what is true. It emerges from the elemental experience of the world as sacred. Mythic narratives are one expression of that aspiration. Scientific narratives are another. Throughout this book we have explored the parallels between science and myth and seen that science retains living roots in myth. Through explicit examples we have seen myth and science illuminating our constant aspiration to more intimately know what we experience as sacred. Science achieves this by providing us with different forms of hierophany, from satellite images of the Earth at night to PET scans of the human brain. Science's hierophanies also come in the form of its narratives that illuminate the night sky, the patterns of human evolution, and the complex workings of Earth's climate. These gateways allow us to see the world as if for the first time and experience the mystery that lives at the heart of human being. In this chapter we have taken our exploration of the constant fire one step further. We have asked a question that penetrates to the core issue of science and religion.

Ultimately we might expect something to exist below our "mere" experience of the world as sacred. We expect there to be the ground, the source, from which truth arises. But is there anything we can directly ask about this truth and its connection to the sacred? What is this truth? Does it exist outside human consciousness as an independent entity? Or is it just a facet of consciousness, ultimately reducible to neurology? The first view would be the hope of the religious advocate. The second would be the claim of the hard-core skeptic. So now we ask is there *the sacred*? Is there a ground from which truth arises, and, if so, what lies on the other side of the "arising"?

In the traditional science and religion debate the usual response to this question has been to ask about God and the powers he/she/it has to ordain, prescribe, or limit the laws of nature. As we have already seen this is not a path we want to follow. As Ursula Goodenough has so eloquently emphasized, if we accept the power and premises of science, then we cannot ask for a supernatural deity that stands outside of nature. Our sense of the sacred must be found woven through and illuminating the universe of natural law. If we accept William James's emphasis on religious experience over theological theorizing, then we must also be careful not to drop the weight of an unsupportable metaphysics on the immediacy of those personal experiences. It is easy to hypothesize the number of angels dancing on the head of the cosmic pin, but it is of little use if no one can attend the performance. Still, with all we have learned we are now in a position to stare directly into the bright sun of these questions and, perhaps, catch something useful in the afterimage burned momentarily onto our imaginations.

In this chapter we have looked at universal, transcendent truths as they are believed to appear in science and myth. On the science side we poked around in the realms of mathematics and its "unreasonable" effectiveness as a description of the physical world. We found that it is the stability of mathematical descriptions of nature across time and space that gives science the appearance of providing unassailable natural law. In mythology we saw deep similarities in sacred narratives across widely different cultures and historical (or prehistorical) periods. This has led some scholars to imagine that myth also carries truths that transcend history and rise to the level of the universal.

Bright parallels run through the discussion of mathematical truths in science and the possibility of transhistorical truth in myth. Both domains hold similar polar responses to the idea of truth. On the one hand, we have the archetypes in myth and the Platonic realm in mathematics. Here the natural law of mathematical sciences and the repeating stories of myth can both be seen deriving their truth from the existence of a transcendent realm. In myth this is the realm of the archetypes and the collective unconscious. In science it is the Platonic realm of ideal forms.

On the other hand, we have the view of structuralism and cognitive science. Mathematical truth and the truth of myth are both expressions of the brain's evolution and the conscious structures that emerged from that evolution. On this view there is no separate truth "out there," no realm of archetypes and Platonic forms. The considerable power of science and myth are to be found in the embodied mind and its response to the physical world, the only one that exists.

So which is it? What do our experiences of the sacred refer to? Do they point us back through the archetypes to a numinous transcendent reality, or does the pervasive and penetrating sense of the sacred stem from the evolution of our neural pathways? The dichotomy here can be cast as polarities — romantic versus rationalist, mushy mystic versus dry realist. It would be easy at this point to drop into the same old dualities that people have been arguing over for centuries. The divisiveness of these dualities belies the complexity and subtleties of what is really going on, if that can ever be known.

Speaking personally, I am a born romantic, but, like Pauli, I am a rationalist too. I am distrustful of metaphysics that cannot be proved or disproved. I do, however, believe in the primacy of direct investigation and direct experience. That is where the sacred appears most convincingly. The sacred appears in people's lives not as a single experience, like a pleasant drug trip, but as an ineffable presence that ebbs and flows. It illuminates and transforms lives. It inspires a deep and committed sense of gratitude, reverence, compassion, and concern. Thus the experience of the sacred is more real than any theories of its location or its ultimate truth.

The sense of profound beauty Bertrand Russell experienced in mathematics is something I and others stretching back to Pythagoras perceive as a hierophany, an expression of the sacred. If pushed I will tilt toward the Platonic (in science at least), but I am not putting my money down on anything other than the depth and power of my own experiences. We must keep asking the question, but, as the singer Iris Dement says, we must, ultimately, be content with the mystery.

In her reflections on the traditional dichotomies of truth in science and religion Ursula Goodenough gives us a coda for our questions about the sacred and the source of truth in science and myth. She writes:

We are all, each one of us, ordained to live out our lives in the context of ultimate questions such as:

Why is there anything at all rather than nothing?
Where do the laws of physics come from?
Why does the Universe seem so strange?

My response to such questions has been to articulate a covenant with mystery. Others of course prefer answers, answers that often include a concept of God. These answers are by definition beliefs because they can neither be proven nor refuted. . . . The opportunity to develop personal beliefs in response to questions of ultimacy, including the active decision to hold no beliefs at all, is central to the human experience. The important part, I believe, is that the questions be openly encountered. To take the Universe on — to ask Why Are Things As They Are — is to generate the foundation for everything else.[34]

The encounter with that great question is the core experience of the sacred; from it the constant fire of our aspiration is lit. Answers are only one part of our need for myth and for science and for religion. In the process of all our storytelling the snake takes its tail, the circle is closed, and we are irrevocably drawn closer to that mystery that begins again every day as we wake anew. If we take this position, then the whole debate about and between science and religion takes a different turn. From this vantage we can change attitudes and assessments and begin to frame the debate in a way that reveals the extraordinary, the miraculous, in just being alive. Ultimately that should be the point of practicing either science or religion or both.

But there is more to the constant fire than issues of sacredness and being. It has a moral imperative as well as an ontological one. We are, as embodied creatures, faced with ethical issues as well as questions about the nature of the True and the Real. In the face of our headlong rush into a future that holds dangerous uncertainty, the connections between science, myth, and the sacred take on urgency. That, finally, is the direction in which we now turn.

CHAPTER 9

A Need Born of Fire

Mythos, Ethos, and Humanity's Most Dangerous Century

Science and technology revolutionize our lives, but memory, tradition and myth frame our response.

Arthur Schlesinger Jr., *The Cycles of American History*

Wild, dark times are rumbling toward us, and the prophet who wishes to write a new apocalypse will have to invent entirely new beasts.

Heinrich Heine, *Lutetia*

Learn therefore O Sisters to distinguish the Eternal Human That walks about among the stones of fire in bliss & woe.

William Blake, *Jerusalem*

Robert Goddard was condemned to see futures that eluded everyone else. He was a born dreamer, but he had the intellect and resolve to patiently work his vision into physical form. It was on the weight of Goddard's shoulders that the great mythic narrative of the twentieth century would be forged. What Goddard dreamed was spaceflight. From the wellspring of his imagination humanity would find the ancient story of the hero's journey transformed into a common endeavor played against the boundless frontier of the stars. But his own journey would be marred by ridicule and neglect.[1]

Goddard's aspiration to spaceflight came at a time when few others had conceived of the possibility. His dream was born on a mild October day in 1899. As the seventeen-year-old climbed into a tree on his family's Worcester, Massachusetts, farm, he saw his future laid out before him.

> It was one of the quiet, colorful afternoons of sheer beauty which we have in October in New England, and as I looked toward the fields at the east, I imagined how wonderful it would be to make some device which had even the possibility of ascending to Mars, and how it would look on a small scale, if sent up from the meadow at my feet.[2]

Reflecting later on that seminal moment, Goddard wrote, "I was a different boy when I descended the tree from when I ascended, for existence at last seemed very purposive."[3]

Goddard was serious about spaceflight, and that meant getting serious about rockets, the only way he could imagine reaching beyond the atmosphere. After obtaining his Ph.D. in physics from Clark University in Worcester, Goddard left home for a research position at Princeton. Ill health, which plagued him much of his life, forced him to return to Worcester and eventually to a professorship at Clark. There he began his rocketry experiments in earnest. Goddard systematically worked through the obstacles, both theoretical and practical, to the creation of powerful, controllable rockets.

Goddard first had to grapple with the dismal efficiency of existing solid fuel "gunpowder" rockets. Only a tiny fraction of the chemical energy locked in the black powder was converted into motion of the missile. To overcome this barrier Goddard began attaching different bell-shaped nozzles to the butt of his rockets. These nozzles allowed more of the heat released in frantic fuel combustion to be directed into thrust, the action/reaction pair that set a rocket in motion. He needed more thrust to push bigger rockets to higher altitudes at higher speed. In time Goddard saw that even with the new nozzles, the solid fuels used since the Han dynasty in China two millennia ago could never drive the thrust needed to bring his vision within his grasp.[4] Beginning in 1919 Goddard began a long series of bold experiments. He hoped to prove that liquid fuels, mixed in a combustion chamber above the bell-shaped nozzle, could propel large rockets on controlled ascents high into the atmosphere. His own ascent through these experiments would, however, come with the high price of derision and public humiliation.

"Aim to Reach Moon with New Rocket," was the front-page *New York Times* headline on January 12, 1920. The story under the banner reported on Goddard's work, recently summarized in a book titled *A Method of Reaching Extreme Altitudes.* Using theoretical arguments and experimental results, Goddard detailed the possibilities of high-altitude flight and conjectured that a rocket might someday reach the dark side of the moon. The *Times* article was respectful in reporting Goddard's findings. The unsigned editorial published the next day was not. It threw respect, good scientific sense, and Robert Goddard's public reputation to the wind.

The January 13 editorial heaped scorn on Goddard's work, including the moon proposal. "After the rocket quits our air and really starts on its longer journey," the editors explained, "it will neither be accelerated nor maintained by the explosion of the charges it then might have left. To claim that it would be is to deny a fundamental law of dynamics, and only Dr. Einstein and his chosen dozen, so few and fit, are licensed to do that." The editors somehow convinced themselves that a rocket could not work in space. Worse, they transferred their ignorance to Goddard. They claimed, "[Goddard] does not know of the relation of action to reaction and the need to have something better than a vacuum against which to react." Then they smacked him down with the lowest insult: "[Goddard] . . . seems to lack the knowledge ladled out daily in high schools."

It was a nightmare. Not only was the editorial wrong on all scientific grounds; it was a personal attack as well. In the wake of this publicity debacle Goddard retreated into the solitude of his work. In that solitude the quiet professor found his way and made his progress. On March 19, 1926, Goddard's first liquid fuel rocket, an ungainly spindle of a creature called Nell, rose 41 feet into the air in a 2.5-second flight. It was a short but momentous journey proving principles that later propelled the space program of every nation on Earth. By the mid-1930s the reclusive Goddard had moved to Roswell, New Mexico, to experiment with larger, more powerful rockets. He continued to build, experiment, and learn. The addition of fins gave his creations stability and control. The use of more powerful liquid fuels shot his rockets to altitudes of 9,000 feet. In his continuing theoretical studies he worked out the principles for gyroscopic guidance and multistage engines. Step-by-step, methodically, Goddard was forging his dream into reality.

Others saw the future taking shape in the desert, but they were not the ones Robert Goddard hoped for. As World War II drew closer Goddard noticed that the German scientists who had occasionally sent him inquiries fell silent. He contacted the U.S. military to explain his concerns and the potential use of liquid fuel rockets for war. After watching a few rocket launches, the visiting generals thanked the professor for his time and left. Five years later thousands of liquid-fueled German V-2 missiles, each carrying one ton of explosives, smashed into London. A new, more sophisticated means for delivering death and terror had been discovered. After the war, when captured German rocket scientists were questioned about their creations, one of them shot back to his interrogator, "Ask your own Dr. Goddard. He knows better than any of us." Goddard had correctly envisioned more than one kind of future for his creation. Later Goddard acknowledged the V-2 as the child of his own dreams but never lost sight of what he hoped would be the true purpose of his rockets — to liberate humanity from gravity's bondage.

Twenty-four years after Goddard's early death in 1945, the world finally came to fully appreciate the man who had brought humanity to the edge of the stars. As Apollo 11's bell-shaped, liquid-fueled rockets locked it on a path to the Moon, the *New York Times* finally bowed to the depth of Goddard's vision. Acknowledging that rockets can indeed fly in space, it retracted its 1920 editorial. "The *Times* acknowledges its mistake," it said. The apology came too late for Robert Goddard but just in time to catch the zenith of humanity's greatest modern myth — the conquest of space.

. . .

The future is a place of myth. As Mircea Eliade pointed out, myth tells stories that answer our most fundamental questions: who we are, where we came from, and where we are intended to go. In many ways the future is the quintessential and, perhaps, only thoroughly modern mythical narrative. As Eliade wrote in *The Myth of Eternal Return*, Paleolithic and Neolithic cultures had no future.[5] Instead they imagined time to be an ever-recurring cycle. Each year the sun charted its path through the sky, animals followed their migratory patterns, fields grew rich with crops and then were harvested. *The Golden Bough*, James Frazier's monumental work on the mythology of agrarian cultures, details a similar conception of time in the mythic cycle of regicide.[6] Each year the king would be killed, symbolically or in bloody reality, to renew the creative energies of the world. For these, our ancestor cultures, time always returned to its starting point. The future would be no different from the past. Average human life spans at this point were short and written records did not exist so what was known of the past was necessarily limited. Change or "progress" was too slow to be fully remembered or understood.

The concept of the future began to change as humanity entered the Axial Age, but even during most of the Christian era the future was limited by the return of Christ and the accompanying apocalypse. Thus the future, a human era that would be notably different from the present, was a creation that paralleled the development and advance of science. From the Renaissance on, knowledge seemed limitless, and so were the possibilities of the human future. In a sense the future was born in the Renaissance, matured in the Enlightenment, and was given full form in the nineteenth and twentieth centuries.

Mythologies of the future, narratives of our potential, have always been mythologies of future science. The braiding of myth and science explored in the previous chapters must now be combined with an understanding of these narratives of the future. In the opening decades of the twenty-first century we have many futures appearing to us, and many of them appear bleak.

Science does more than function as myth in a global culture saturated with its fruits. It does more than simply recapture the great themes of myth. Human society's vision for itself, its mythos of the future, cannot be separated from science. Our narratives of the future are shaped in contexts of what science makes possible, or makes impossible to escape. From shiny utopian futures of peace and cosmic exploration to nightmares of nuclear or environmental apocalypse, our stories of the future are myths. They are myths born from imaginations made fecund by the reach and vision of science. In this chapter we consider the human future and the ethical imperatives that follow in the wake of weaving together science, myth, and the sacred.

Throughout this book I have built the case that science must be recognized as a hierophany, a gateway to the experience of the world as sacred. Making this leap is not simply a matter of changing the academic debate between science and religion. Finding the appropriate resonance between our scientific and spiritual endeavors calls us to develop a new ethos, a planetary ethos, as well.

Isaac Asimov, the great master of science fiction, once looked at the world's problems and drew exactly the wrong conclusion. "The dangers that face the world," he wrote, "can, every one of them, be traced back to science. The salvations that may save the world will, every one of them, be traced back to science."[7] Asimov got it wrong because Logos, the purely rational and analytic vision, is but one human response to the world. Mythos, the need for narratives that express the experience of the world's sacred character, is the other. As we have seen Mythos and Logos cannot be neatly separated. To act collectively humans must have a context in which to set their actions. Recognizing science as hierophany allows it to be brought into the embrace of the deepest human values that emanate from an experience of sacredness — compassion, care, and gratitude. The pairing of science and myth gives us a fulcrum on which to raise a different understanding of science and its application. One means to appreciate this is to explore how science now provides a potent mythic system for narrating the most precious of all imaginative resources — our collective future. We begin where Robert Goddard left us, the high frontier and the conquest of space.

FROM ROCKETMEN TO CYBERPUNKS: MYTHIC FUTURES REAL AND IMAGINED

> Space . . . the Final Frontier. These are the voyages of the starship *Enterprise*. Its five-year mission: to explore strange new worlds, to seek out new life and new civilizations, to boldly go where no man has gone before.

It's hard to find anyone in the United States who doesn't know at least some part of *Star Trek*'s opening lines. My eight-year-old son had never seen a single episode of the series but still managed to toss out the last line over hamburgers one day when we were discussing space travel. First aired in 1967 during the height of the space race, *Star Trek* epitomized the dream of a quasi-utopian society born in the wake of our conquest of space. In sense and sentiment the implacable Captain Kirk's words echo that of President John F. Kennedy when, at the height of the Cold War, he launched the Apollo program and our march to the Moon:

> We set sail on this new sea because there is new knowledge to be gained, and new rights to be won, and they must be won and used for the progress of all people. . . . We choose to go to the moon. We choose to go to the moon in this decade and do the other

things, not because they are easy, but because they are hard, because that goal will serve to organize and measure the best of our energies and skills, because that challenge is one that we are willing to accept, one we are unwilling to postpone, and one which we intend to win.[8]

The "conquest of space" is a story with two sides that have played off each other since Goddard's time. On one side is the scientific-technological reality of astronautical engineering, hardware, mission plans, and budgetary constraints. On the other side are the popular science fiction narratives that have extrapolated the hardware to imagine what we might become in worlds bequeathed to us by ingenuity and determination. If the future is a place of myth we can learn something of its recent, accelerated evolution through the intertwined stories of space exploration and the possibilities imagined in science fiction. Together these narratives allow us to see where the interplay between myth and reality has left us at the beginning of the twenty-first century. Together they help us understand the boundaries of our own choices as we face the difficult times ahead.

President Kennedy's commitment to a Moon landing came in the wake of successful Russian efforts to reach earth orbit. *Sputnik*, the first artificial satellite, was launched by the USSR in 1957. Its incessant radio bleeping from five hundred miles overhead sent shock waves through the U.S. political, military, and scientific establishments. The conquest of space became entangled in terrestrial battles for geopolitical supremacy. But when Kennedy set the United States on the path to a lunar landing he used language that called to something higher, a language of purpose and possibilities. Kennedy's words called on the myth of a future in space, a myth that had been taking shape for years. It was a culmination of an ageless dream to know the stars and the more recent century of scientific experimentation in service of that dream. Like Goddard's reverie in his family's cherry tree, Kennedy's vision for the nation's space program would be driven into form through intelligence and perseverance. Unlike Goddard's quiet, solitary efforts,

Kennedy's vision would require billions of dollars and the greatest scientific establishment in history.

For a decade the United States committed itself fully to placing humans on the surface of the Moon. While the Soviet Union continued breaking new boundaries in human and robotic space travel NASA became the focus of our own nation's effort. Using rocket technology pioneered by Robert Goddard and refined by German scientists during World War II, NASA scientists and engineers overcame one hurdle after another while the nation, and the world, watched.

We started into space single file and then quickly widened the line. First came the Mercury program. There were seven Mercury launches between 1960 and 1963. Solitary astronauts were packed into a capsule smaller than a VW bug and blown into orbit atop converted intercontinental nuclear missiles. Each launch taught NASA more about the critical issues of spacecraft control and the dangers of reentry. After Mercury came the two-man Gemini program. Eight Gemini launches (once again on converted nuclear missiles) where carried out between 1964 and 1967. With the Gemini series astronauts learned how to leave the safety of their capsule and how to rendezvous with other orbiting spacecraft, both critical skills needed to get humans to the Moon. Finally came the three-man Apollo program. To reach the Moon repurposed nuclear missiles would not do. NASA responded with its towering thirty-two-story, three-stage Saturn V rocket. Its enormous power was needed to drive a tiny spacecraft out of Earth's deep well of gravity and drop it into orbit around the Moon.

Every mission from the first Mercury launch to the final Apollo splash-down had its dangers. The press carried news of each launch and the astronauts' progress in tense prose that captured the sense of wonder and anxiety. The U.S. space program quickly accumulated a list of astronaut heroes, Jason and Odysseus in spacesuits. A few of these heroes died in pursuit of the great purpose, but that seemed to be the price for opening up such a vast and limitless frontier. This was the stuff of dreams after all. The space program was a collective odyssey, a matrix of vision, ideas, and values that feeds and creates myth. When Apollo 11 finally touched down on the Moon the entire world was watching. People understood, if only dimly, that we had passed across a threshold.

Along the way to the Moon, NASA also pushed its robotic space probes out to the planets. The Mariner program navigated three billion miles of interplanetary space and managed to get six of nine probes to Mars, sending back stunning images of a desert red planet. Mercury and Venus were also targeted and achieved. The United States was not alone in its exploration of the high frontier. Russia was sending its own flotilla of spacecraft to the planets with various degrees of success. All at once, after millennia of dreaming about the planets, humanity was sending its proxies to visit and take tourist snapshots. How could anyone at the time doubt that we lived in a world of wonder, a world of boundless possibilities made real through the capacities of science?

On December 19, 1972, Apollo 17 lifted off the launchpad on a pillar of flame. As the crushing weight of acceleration pushed the three astronauts into their seats neither they nor anyone else could guess that this would be humanity's last trip to the Moon. Just two and a half years after Neil Armstrong's small but epoch-making step onto the lunar surface, Apollo would be canceled for lack of interest. Three more missions had been planned. They were all sacked as Congress and the president contended with a failed war in Vietnam and domestic unrest at home. The times were changing.

In the three decades that followed Apollo, NASA continued to achieve stunning successes in the domain of robotic exploration. Jupiter and Saturn were explored from orbit. Mars was visited by a fleet of spacecraft that carried robotic landers. The Hubble space telescope and other orbiting astronomical platforms opened unimagined windows onto the cosmos. These were all tremendous and lasting achievements. They may, in fact, be our nation's most important contributions to science. Still, something seemed missing. For many observers the conquest of space by robots, irregularly shaped boxes of electronics with solar panels, seemed less than what the great visionaries of the twentieth century had in mind. In the decades since Apollo something seemed lost. The dreams of lunar colonies, Mars expeditions, and a burgeoning interplanetary culture waiting just ahead of us were deferred to a later date.

It was in the arena of manned spaceflight that the vision stalled. After Apollo, NASA and the other space agencies focused not on the conquest of space but on the exploitation of near-earth orbit. NASA's manned effort went into the Space Shuttle and the International Space Station. Both these projects were ambitious technologically but lacked the clear focus and larger vision of Apollo. In the case of the Space Station it was hard for scientists and the public to understand what the expensive, decade-long project was built for, other than doing it "because we can" and because it gave contracts to aerospace corporations. A growing sense of drift enveloped the manned space program throughout the 1990s. When the Shuttle *Columbia* exploded on reentry in 2002, killing all on board, it was on a mission whose scientific objectives included "mixing paint with urine in zero-gravity, observing ant farms, and other comparable activities — all done at a cost greater than the annual federal budgets for fusion energy research and pancreatic cancer research, combined."[9]

The loss of an ambitious vision for human space exploration can be seen as the loss of a greater myth for the future. It is no surprise that the cancellation of Apollo paralleled a change in the stories of science fiction. At about the same time the manned space program stalled, our narratives of possibilities began to shift their emphasis. Science fiction always responded to the realities around it, and we can chart a similar arc from an ever-expanding future to a diminishing sense of constraint in the midst of high technology. We begin with the rising slope of the arc and science fiction's vision of the conquest of space.

The USS *Enterprise* of *Star Trek* and the giant Saturn V rockets of the Apollo program were born of the same mythic vision of who we were and what we might become. In its particular manifestation, *Star*

Trek was a creation unique to the United States and to Western civilization in its long trek from the first Greek scientists to the dreams of Enlightenment rationalists. It doesn't take a Ph.D. in cultural studies to notice how much *Star Trek*'s United Federation of Planets looks like a United States of the Universe. Still, its grand imagining of a united human future free of (internal) warfare and full of endless exploration was compelling. It captured the optimism embodied by the Apollo program. The show was canceled after a few seasons, but it would go on to become iconic, the perfect representation of the grandest vision of what science might allow us to become.

Star Trek was, of course, not the only dream of a boundless future that awaited us as scientific advances opened the doors to the conquest of space. In fact, it was a culmination of that dream. Throughout the twentieth century science fiction books and movies charted the landscapes of imagination, and many of its mappings showed us worlds of pure promise. In the wake of World War II and the stunning advance of science, the fiction in these science fiction stories began to seem less insular, less the domain of geeks, and more part of a collective cultural dream. The pervasiveness of its images in popular culture made it into a shared myth.

The classic *Foundation Trilogy* by Isaac Asimov was a good example of postwar optimism about what the human expansion of space would mean. Asimov's trilogy was published in 1949 and went on to sell millions of copies and spawn a continuing series of sequels and prequels. In Asimov's future the Milky Way Galaxy has been entirely colonized by humanity. A billion inhabited worlds support a human population of 100 trillion souls. It is not a utopia. The cycles of civilization, with their rise and fall, operate even in galaxy-spanning cultures. In spite of these "realities" Asimov had captured the essence of the myth of a boundless future by presenting readers with a vision of nearly infinite human expansion. We would continue forever. Other writers of this "classic" era of science fiction had their own take on space travel and the future, but the basic optimism was often the same. Movies followed suit. The famous 1956 film *Forbidden Planet*, based loosely on Shakespeare's *The Tempest*, created a perfect pulp fiction universe with a united human civilization (which seemed to consist only of white people) colonizing innumerable worlds.

What tied these fictional universes together was their background of shiny omnipotent technology. The main characters lived on massive, powerful starships. Every ship had its crew of technicians standing before banks of blinking lights in perfect control of machines that could navigate the void or shape worlds. The universe still contained dangers in these stories (often driven by our own inescapable flaws), but what mattered for the myth of the future was that humanity had left its nest. We had become something more, something greater.

This essential optimism did not exist alone, of course. Other writers and artists could feel the pull of the shadow. The mythologies of the future had always understood darkness. Even as Asimov and others dreamed of gleaming techno-topias others were more concerned with the destructive powers science unleashed. Stories of bug-eyed monsters and evil aliens have always existed in science fiction, but these were simply the demons of our previous mythologies given new costumes. After atomic bombs vaporized Hiroshima and Nagasaki, however, a cloud that had not existed before passed over the future. Like the classic film *Godzilla*, countless science fiction movies and stories of the 1950s provided meditations on the powers we had let loose and were not prepared to control. Here a mythological antecedent did exist. When Robert Oppenheimer saw the mushroom cloud rise over the Trinity test site in New Mexico he quoted the most famous of Hindu scriptures, the Bhagavad Gita: "Now I have become death, the destroyer of worlds."[10] Apocalypse, the end of all time, had always been the domain of myth. Science had given us the means to see what it might look like in reality.

Throughout the 1960s a kind of standoff had been achieved between visions of a boundless space-faring future and the apocalyptic nightmares of nuclear war. Even with the grave and ever-present danger of nuclear conflict, the culture of post–World War II America and Kennedy had been one of hope. Apollo had been the most concrete manifestation of that optimism about where our future might take us. By the early 1970s that hope had begun to fade and with it new, more claustrophobic visions of the future began to emerge. In time these would come to dominate the science fiction landscape. Shiny utopias gave way to the dirty futures of dystopia.

The first Earth Day was held on April 1, 1970. Eight years earlier Rachel Carson had published *Silent Spring*, a warning that pollution and pesticides were damaging the natural world. The first visions of environmental degradation entered cultural consciousness just as Apollo was winding down. Other threads were then woven into a changing vision of the future. During the 1970s the United States experienced two oil shocks. For the first time people confronted the possibility that the fossil fuel party they had been attending for seven or eight decades might not continue indefinitely. The loss of U.S. industrial prowess to Japan, Germany, and others and the humiliation of the Iran hostage crisis drove hope into cynicism. In spite of renewed economic growth in the 1980s and 1990s concerns about environmental degradation continued, with global warming beginning to impinge on popular consciousness. Even the engines of growth in the 1980s and 1990s drove fears that were manifested in science fiction's vision. While globalization, biotechnology, and the rise of computer networks allowed some to imagine new futures of unlimited frontiers, others saw something far more forbidding. In popular consciousness it seemed that something had turned. The most pervasive visions of the future seemed stuck in an ever-descending spiral.

One particularly influential vision emerged in the 1980s in a form called cyberpunk.[11] Beginning in 1984, William Gibson, an American expatriate living in Vancouver, became the genre's founder with a series of books both visionary and prophetic. It was Gibson who single-handedly invented key features of the terminology and imagery we take for granted in our real Web-laden culture. *Cyberspace* was his term and his invention. What he saw was a dirty future, a Darwinian high-tech nightmare overrun by global corporate elites wielding more power than nations. Below them moved the vast masses feeding at the bottom of an information economy that had morphed into a kind of transcendent alternate reality. Gibson imagined a claustrophobic dystopia of human beings modifying their bodies in the service of technology and endless sprawling cities where decay mixed with hyper-tech innovation.

The universe of cyberpunk has been given many forms by authors such as Bruce Sterling (considered cofounder of the genre with Gibson) and John Shirley. Cyberpunk themes have also been explored by non–science fiction writers. The feminist author Marge Peircy's 1991 book, *He She It*, combined themes of environmental destruction with the degradation of human life under a corporate-dominated information economy. In film, critically acclaimed small films as well as box office mega-hits, the cyberpunk ethos and vision became dominant. *Blade Runner* predated Gibson's first book, *Neuromancer*, by two years but was later hailed as capturing the essential, dark vision of cyberpunk. Most notable and noteworthy were the wildly successful *Matrix* films. In the universe of the Matrix human beings have lost a war with artificial intelligence, machines of our own creation. In the ultimate act of environmental destruction humans "scorch the sky" to deny machines solar power. In response, the victorious machines reduce humanity to a power source. Human beings become mere bio-batteries, made docile through enforced connection to a virtual world that re-creates the late twentieth century. The Matrix takes dystopia and the human/computer interface to extremes. The radical popularity of the *Matrix* films (the series took in more than one billion dollars) is testimony to the pervasive influence of the cyberpunk vision. As a whole, the myth of the future had gone to the dark side.

It is worth noting that space travel is often not an essential part of the cyberpunk vision. Like the real world we live in, the conquest of space stalled in these stories. Space travel was confined to near-earth orbit or to outposts of industrial exploitation that seem hellish and far away. In some stories nations have completely given up on their space programs, and it is either large corporations or smaller private groups that push forward. In cyberpunk the grand vision of exploration in space has been, for the most part, replaced with a darker vision of survival on Earth.

Of course, science fiction forms a diverse set of voices. Some people who are familiar with its forms could take issue with the sketch I have outlined here. Along with the darker vision of cyberpunk there were writers and directors exploring more optimistic and hopeful futures. *Star Trek* itself gained its popularity in the mid-1970s, years *after* its cancellation. Still, it is difficult to argue with the fact that in the last thirty years of the twentieth century the mood of our future visions has changed. That change paralleled a loss of the mythic narrative that accompanied the conquest of space. Now, at the beginning of the twenty-first century, we find ourselves facing a very different world and a very different sense of the future. Those who oversaw the heady acceleration of science in the previous century had reason to see the future as boundless. Now we face boundaries that cannot be escaped. The change in the myth of the future is testimony to the greater change in our understanding of what science is as a cultural force. More important, the shift in our myths of the future allow us to see constraints that science has placed on that future.

THROUGH THE BOTTLENECK: A CENTURY ON THE BRINK

It was only my second month working at the Goddard Institute for Space Studies (GISS), and I still was clueless about climate science. I got the job in November 1985, three years before James Hansen, the director of GISS, would make headlines and history with his congressional testimony on global warming. I was a newly minted physics student with a B.S. degree, taking time off before I started work on a Ph.D. For a year I had grabbed any job I could find, including planting trees in

British Columbian clear-cuts (deforestation up close and personal) and an unlikely gig as a bouncer at the *Rocky Horror Picture Show* on Eighth Street ("Hey, you in the bathroom, stop that!"). After those strange detours I was hoping to finally get a job doing something, anything, in science. The position as assistant computer programmer at GISS was a dream come true.

Every day for five weeks I had taken the subway up to 113th street and GISS's cramped offices. There in a tiny windowless office amid stacks of old printouts, I reveled in learning about Fourier transforms, autocorrelations, and basic programming skills. The mathematics was beautiful, but I was still not sure what I would be applying it to. I decided to ask Inez Fung, one of my bosses, if she could explain a little more about this whole climate science thing. Inez was always upbeat and patient with me. Her kindness made it easy to ask what seemed like endless stupid questions. Astronomy and quantum physics had been my great passions in undergraduate work; climate science seemed pedestrian by comparison. Two hours later I staggered out of Inez's office and left the building for a long walk. The only thing pedestrian that day would be me.

I had never heard of global warming or climate change. I had never heard of melting glaciers, the breakup of ice sheets, or rising sea levels. It was all new. Inez laid it out for me as best she and her colleagues understood it back then. Their computer models, chugging away on massive machines taking up most of the third floor, were crude. But in spite of their low resolution and incomplete physics, the GISS climate models were already outlining the basic predictions we are so familiar with now. The Earth would get warmer on average. More energy would be dumped into the climate system. The system, in response, would change. The change might be dramatic and might pose serious threats to human populations.

I was an impressionable twenty-four-year-old. The story Inez had crafted impromptu with graphs, graphics, and maps lying around her office left me shaken. I wandered around the streets of Manhattan for hours. The traffic buzzed. Oceans of people moved through the streets. Suddenly nothing of the life around me could be taken for granted. I made my way down to Battery Park at the southern tip of the island. The ferries docked, left again, and returned. The towers of Wall Street rose above me. I sat and imagined the whole scene underwater. How could it be possible? Manhattan lost to rising oceans? No way. But I had just learned that it was possible and might even be probable unless things changed. "How long?" I asked myself. "How long will this city, this life, continue as it was, all around me? Will it get bad? What will we all do if it does get bad?"

For months afterward I tried to explain global warming to my distinctly nonscientist friends. Over beers at a bar in the East Village they would listen and nod. "Whoa, dude," they would say. "That's gnarly." They didn't believe me. I didn't believe me. This was the mid-1980s. Nobody knew what I was talking about, and I knew there was no way to really tell anyone. It was too big, too abstract, too impossible to fathom. Now, twenty years later, climate change dominates the news. The headline of the week is Canada's largest ice sheet coming apart at the edges. Two decades later climate change hardly has to be explained. Now it is seems like something coming from firsthand experience. Now it is seems like something we are going to have to deal with.

Climate change is not our only problem. What makes our moment in history so sobering is the banal fact that global warming's enormity comes as but one of a series of interlocking challenges facing humanity and our project of civilization. These challenges will bear down hard on us over the next century.

In his 2002 book, *Our Final Hour*, Sir Martin Rees reviewed a variety of threats facing the human race that extend beyond global warming and environmental destruction. Of particular concern for Rees is the specter of nuclear and biological terrorism. The addition of North Korea and perhaps Iran to the nuclear club is one cause for concern, but the creation of small atomic weapons or, at least, radiological dirty bombs is an increasing possibility that is more disturbing for Rees.

Such concerns do not threaten the world, of course, but they pose the risk of making large, crucial cities uninhabitable. Of greater impact and greater concern for Rees is the possibility of lethal bio-weapons. The most potent danger in his eyes is the development of powerful biological agents by small groups or even disgruntled individuals. The ability to create and release such agents is becoming dangerously decentralized. Rees quotes former Undersecretary of Defense Fred Ilke:

> The knowledge and techniques for making biological super-weapons will become dispersed among hospitals, agricultural research institutes and peaceful factories everywhere. Only an oppressive police state could assure total government control over such novel tools for mass destruction.[12]

This may seem like the stuff of science fiction, but Rees stands behind his claims. In 2003 he registered a wager with the Long Bet Foundation that by 2020, bioterror or bioerror will lead to one million casualties in a single event.

Rees considers the variety of dangers we face, including environmental ones, and sums their total impact to measure our chances of survival. He gives humanity 50/50 odds of making it, intact, to the next century. It is a sobering calculation. His interest in high-tech threats such as genetically engineered superviruses and out-of-control self-replicating nanobots reflects his vision as a scientist and his understanding that science fiction turns into science reality very rapidly. Some may find Rees's emphasis on bioterror and nanobots overreaching but the idea that we face a unique and uniquely critical epoch in human history has already become mainstream.

Scientific American is hardly a shrill voice of extremes. In September 2005 it dedicated an entire issue to "Earth at the Crossroads." As the magazine's editor, George Musser, put it, "Three current, intertwined transitions, economic, demographic and environmental[,] . . . are transforming everything from geopolitics to the structure of the family. And they pose problems on a scale that human beings have little experience with." Musser goes on to quote the Harvard University biologist E. O. Wilson and his claim that we are about to pass through a "bottleneck," a time of maximum stress on natural resources and human ingenuity. The bottleneck is the recurrent theme in Musser's essay and the other articles in the issue. Humanity now faces dangers unlike any that have appeared before in our evolution. Those dangers can be managed if we act wisely. As Musser puts it, "A bottleneck may be tough to squeeze through, but once you do the worst is behind you."[13] The problem, of course, is that sometimes you don't make it through the bottleneck.

Failure of our entire global civilization is, unfortunately, a real option. History and prehistory offer spectacular examples of societies imploding as a result of their blind choices. This was the topic of Jared Diamond's best-selling book *Collapse.* The question he took on was simple: Why do some societies rise to prosperity and then collapse disastrously when others manage to survive intact for millennia? Diamond, a professor at UCLA, is a prolific writer and scientist. In *Collapse* he explored the issues that faced failed civilizations ranging from Easter Island and the Maya to the Norse colonies on Greenland. Understanding that we face our own precipice, Diamond performed a comparative analysis in the hope of learning what factors and what choices determine success or failure. In all the cases he studied he found that environmental factors, in the form of resource depletion, played a critical role. Human beings, it seems, have a prodigious ability to destroy the very fundamentals on which their societies depend. Easter Island, a barren sixty-four-square-mile outpost in the Pacific, provides one of the clearest examples of societal collapse.

Easter Island is famous for its enormous and enigmatic statues. The size of the statues (up to 90 tons) and the fact that the island is completely barren of trees to transport them (by means of rolling logs) puzzled researchers for many years. In the 1970s the author Erich von Daniken went off the deep end and suggested that aliens had constructed the statues after becoming stranded on the island. The truth, it turns out, has more to do with human fallibility than ancient astronauts.

Easter Island once had trees — lots of them. When seafaring Polynesian colonists reached the island sometime in the first millennium they found all the resources needed to create a rich and prosperous culture. Their religious and political structure included carving the great statues from quarries in an extinct volcano at the island's center. Logs and rope made from felled trees were then used to transport the stone totems to the various tribal settlements. Each clan saw the large statues as testaments to their power and wealth. A statue building arms race ensued. The need for trees, rope, and food to maintain a population of laborers eventually led to the destruction of the very forests the islanders depended on. After the forests were gone erosion took the soil too. What followed was Easter Island collapsing into starvation, warfare, and cannibalism. The chance of escape disappeared too as seafaring canoes require large trees for their hulls. Eventually the statues themselves fell victim to the collapse. Warring tribes pulled down their enemy's statues as a permanent night descended on Easter Island's society. What remained was only a haunting legacy of ruin.

"What was the guy who cut down the last tree on Easter Island thinking?" one of Diamond's students asked him in class. It is the same question we must ask ourselves as we rush headlong into the bottleneck. Diamond's review of societies that failed and those that succeeded shows that each faced similar problems. Each ran up against limits in natural resources. Each hit ceilings on their available energy, polluted existing resources, and overextended their population. The failures were almost always self-imposed, and in their sad stories Diamond finds compelling commonality. First came failures of vision as the societies were unable to anticipate approaching problems. Once those problems arrived the doomed cultures were unable to perceive the existence of a threat. The final failure came when the challenges were recognized. The societies were unable to change values or modify behavior to save themselves. In many respects the last failure, the challenge of values, may be the most important. As Diamond writes:

Perhaps a crux of success or failure as a society is knowing which
core values to hold on to and which ones to discard and replace with
new values. . . . Societies and individuals that succeed may be those
that have the courage to take those difficult decisions and the luck
to win their gambles.[14]

Core values are not simply rational choices; they are also the domain
of myth and spirit. Our values emanate from what we believe has meaning, what we take to be of value. In a word, our values derive from what
we hold sacred.

On their way around the Moon in 1968, Apollo 8 astronauts snapped
what may be the most important picture ever taken. Their image of the
blue Earth rising above the Moon, like a marble against the blackness
of space, was epoch making. For the first time the whole planet could be
seen for exactly what it is — an isolated outpost in the vacuum of space.
We are a planetary Easter Island. There is nowhere to escape to and,
perhaps, not much time to figure out how to manage the global cache
of air, water, forests, and croplands. Our success or failure will depend
on what we value, our inventiveness, and our ability to act in a unified
fashion. Arthur Schlesinger Jr. saw the problem clearly in the epigraph
at the beginning of this chapter. Our salvation cannot and will not rest
with technology alone. Instead our success will turn on a mix of myth
and science brought to fruition through the application of wisdom.

THE COSMIC TEENAGER
AND AN ETHIC OF INQUIRY

But [science] might yet be found to be the great river of
pain! — And then its counterforce might at the same time be
found: its immense capacity for letting new galaxies of joy
flare up!

Friedrich Wilhelm Nietzsche, *The Gay Science*

Myth serves many functions. One of its critical roles in early cultures
was to provide meaning and guidance during times of transition. Boys

were initiated into the sacred narratives of the hunt as they made their transition into men. A different set of stories accompanied girls in their rites of passage through the shock of first menses and the transition into the domain of women. Myth has always connected people with their deeper sources of wisdom as they navigate the uncertainty of change. By serving as a connection to the sacred, myth transformed the terrifying experiences of transition into experiences of power and compassion.

Humanity as a whole now faces such a transition. In a very real sense we have become, as a species, a kind of planetary adolescent. After a long period of childhood we have grown up. We now have some power over ourselves and our environment. In a sense we have reached the age when we can drive and we have been given the keys to the planet. The decision we face, as a species, is the one that all teenagers face. Will we drive like idiots, getting drunk and racing the world around the edges of sustainability, or will we get a grip and see the responsibility that comes with power. This is a bottleneck we all understand. It is a useful allegory for the other bottleneck that lies ahead, the one we have a hard time collectively wrapping our minds around because it stretches decades ahead of us and involves complex issues of atmospheric chemistry, natural resource depletion, population control, and energy systems.

Given the magnitude of the difficulty ahead we can look at our coming-of-age as a species and ask if there is anything useful in myth that can guide us and provide perspective, illumination, and wisdom. Because we have never before been required to view ourselves as a species, the answer to this question cannot simply be a return to mythological narratives of our ancestors. What we need are guides relevant to a science-saturated culture as we navigate our own dangerous transition. We require wisdom that both reaches back to the sacred narratives of early cultures and stretches forward to grasp the engine of future transformation. This can only be done in the interwoven imaginative domains of science and myth. Throughout this book I have built the case that traditional debates between science and religion have missed the point for a long time. In this chapter I argue that time is not on our side and forging a different path is now imperative.

Recall that traditional theology-based skirmishes in science and religion pit scriptural concepts of divinity against the specific forms of cosmology and evolution. More recently, "Eastern" religious worldviews have been pushed into the debate through claims of their confirmation at the frontiers of quantum physics. We have seen that both approaches are fundamentally flawed. Both approaches miss a simpler, deeper, and more fecund truth. Scientific results do not provide science's connection with the domains of human spiritual endeavor. Instead it is the aspiration of science that matters. That aspiration comes from direct experience of the world's deepest beauty and significance. That aspiration, our constant fire, emanates from experience of what has long been called sacred. It is ancient and was first given expression in the realms of myth. Most important, within myth we find braided threads that became the modern domains of science and religion.

Myth has never left us. It never can. Its emphasis on narrative allows it to frame our meanings and actions. That vital link between meaning and action spurs us to look carefully at the continuing, living connection between science, myth, and the sacred. There we may find the much-needed narratives that can shape a planetary response to our planetary coming-of-age.

How can the recognition of science's living roots in myth and its capacity to act as a hierophany help guide us through the bottleneck? As Diamond points out, the core values we hold are the key to our fate. Thus the emphasis shifts from mythology alone to the mythologies of spiritual endeavor and spiritual practice. If mythologies provide a guide for transitions, then no transition is more difficult than the ascension to some kind of understanding. Like Odin on his path to Mimir's well, we must pass from knowledge to wisdom. To make this transition we must see that science already provides a deep vision of behavior that can form an ethical complement to spiritual endeavor. It is a vision that can guide the wise application of science.

Erwin Schrödinger, a founder of quantum mechanics, once wrote, "The scientist only imposes two things, namely truth and sincerity, . . . upon himself and other scientists."[15] This is true in science, and it is true in all authentic spiritual endeavors. While many skeptics argue that religion serves only as a means of controlling populations, there is no doubt that it has also channeled innumerable personal searches for personal truths. When these have been guided with wisdom and integrity they have forced seekers to stay close to what is experienced. In this sense there is a deep parallel between spiritual and scientific practice.

In 1991 two British astronomers, A. Lyne and M. Bailes, created an uproar when they announced the discovery of a planet orbiting the neutron star PSR1829–10, a dead cinder of a once-massive sun. The result thrilled and shocked the astronomical community. For two and a half thousand years philosophers and astronomers asked if planets existed outside our solar system. Giordano Bruno's execution formed one part of this long story. For two and a half thousand years the question remained steadfastly unanswerable. Lyne and Bailes's discovery seemed to provide an answer. It was big news. Then, a year later, at an astronomical meeting designed to present new results, Lyne stood before a large audience and announced that he and Bailes had it wrong. With news cameras rolling, Lyne detailed how their analysis of the data had been in error. They were withdrawing their claim of discovery. There was a long pause. Then the audience came to its feet in a standing ovation.

Some argue that science is amoral, that no inherent ethical conclusions can be drawn from scientific findings. There is, however, one precept that we scientists take as holy from the time we begin as graduate students: "Tell the truth." There is no greater sin in science than to falsify data or conclusions. Scientists are asked to let the world speak for itself, to observe without bias or preconceived ideas. In the ideal, scientists are asked to witness the world in its own great pathways of beauty without the filter of prior desires or demands. Brutal honesty about the character of the conclusions drawn in the investigations is a hallmark of sincere scientific practice. The scientist has to be honest with himself about the integrity of the result and the possibility of error. That is why the audience saw Lyne and Bailes as heroes to be honored, not as failures to be shunned. Their narrative becomes part of the Mythos of science by calling its practitioners to a set of core values that includes absolute honesty in letting the world speak on its own.

The ethic of honesty and integrity in the investigation of physical reality forms another living parallel between scientific practice and the mythologies of spiritual practice. Again the aspiration to know the world means one must adhere to the ways it presents itself to us. Any understanding of the world's truth must be approached with great effort and the willingness to accept the path that investigation demands. Scientific investigation is hard work and demands honesty and the willingness to shy away from easy answers. Spiritual endeavor as described by writers in all the world's traditions seems to make the same demands. Thus we find an ideal of effort and unflinching honesty that can bind us together in developing an ethic for the judicious and wise application of science. The scholar of religion Huston Smith claims one can evaluate the depth of a person's attainment in any religious practice by the change it brings in his or her behavior, the way he or she lives. An enlarged circle of compassion is a hallmark of such change. The fruit of recognizing parallels in science and spiritual endeavor may be the communal application of what we might call an "ethic of investigation." As a society we must be willing to see our use of science's power with absolute clarity if we are to change our collective behavior.

We must face the vast and challenging truth of issues such as climate change, truths revealed through scientific practice, for what they are and as they are. There is no place for us to hide from their reality, and reality should be the hallmark of both sincere scientific and spiritual endeavor. Once those truths are acknowledged we then have choices in the way we respond. Compassionate, determined action that flows from spiritual practice, as embodied by Gandhi, Martin Luther King Jr., and countless others, is one possible response.

The ethic of investigation is just one possible way in which the weaving of science, myth, and the sacred can inform our choices. Unpacking the relevance of taking science as a hierophany is a project that will take time and require its own investigation. But we can acknowledge that the time is ripe and we are called to action by the path our evolution has taken. We live in a marketplace, a "real world," saturated with the fruits and poisons of science. There can be no doubt that in spite of the grandeur and blessing of its fruits, science's poisons threaten both the human habitability of the planet and our project of civilization. An enlivened perspective on science, myth, and the sacred that focuses on aspiration rather than results and the ethic of inquiry rather than competing claims of truth can provide the much-needed reenvisioning of our relation to the physical world. It can form a landmark from which we sight a stable, sustainable human future on the planet that extends far to horizons we have yet to imagine.

Religion, as an institution, often ends up as simply another form of humanity's never-ending engagement with greed, anger, and delusion. At the same time religious experience can often be the means for profound insight and transformation. Insight and transformation are also hallmarks of science. If we recognize science as a means of manifesting the world's sacred character we can no longer see it as simply a means to an end. If we see that science emanates from the mythic call to create meaning we can no longer make it simply a tool for control of the natural world. If we locate science alongside the field of human spiritual aspiration it becomes an instrument for gaining both knowledge and wisdom. That is the fundamental change. When science and myth are drawn into a parallel complementarity we can see both as a way to affirm life and its place in the cosmos.

Epilogue

Fire in the Open Mind

This is thy hour, O Soul, thy free flight into the wordless,
Away from books, away from art, the day erased, the lesson
    done,
Thee fully forth emerging, silent, gazing, pondering the
    themes thou lovest best,
Night, sleep, death, and the stars.

Walt Whitman

SCENE 1: "Oh, no. I'm in trouble now." The guy standing up looks pretty angry. I brace myself.

I just finished a talk titled "The Constant Fire" at a local university. It was the first time I presented my thinking on science and religion before a scientific audience. Actually, it was the first time I admitted that I think about this kind of thing at all. I'm still feeling pretty nervous. I don't know what to expect exactly, but I have an intuition that some hostility may be involved. Turns out I have good intuition.

It seemed like the talk, with its PowerPoint images of Kepler, William James, Eliade, and Wolfgang Pauli, has gone over pretty well. No tomatoes hurled from the audience yet. Now it's the question-and-answer period, and the trouble starts right away. The man standing before me

255

is, I know, from the biostatistics faculty. He is older and has the air of a stern but knowledgeable teacher. He clears his throat and begins by questioning my assumption that there's any need, or importance, to find proper dialogue between science and religion. He claims that there is really nothing to say about them together because there is nothing that can be defined properly. It is, in his opinion, nothing but wasted hot air. As he speaks his tone becomes more strident, angrier. Finally he shakes his finger at me and, articulating each word slowly, says, "I don't know where you're going with all this, but I will tell you that I do not like it!"

SCENE II: The gentleman standing at the podium across from mine *seems* like a nice guy. He is earnest and polite, a practicing Ph.D. astronomer who studies how gamma rays are emitted by distant exploding objects. He is also a born-again Christian and an advocate of Creation science. That is why we are both here. I watch as he explains to the audience, in some detail, how Big Bang cosmology serves as proof of the biblical story of Genesis. Yikes! Now I'm the one who is angry.

I'm on this stage because someone got hold of my name as a "scientist who thinks about religion." I knew when he called me that this "public debate" on science and religion was really going to be people speaking past each other about Science versus Creationism. Normally, I stay away from such affairs, but in this case I felt the pull of responsibility. On the "Religion" side of the debate there was, of course, the usual antievolutionist. He's a biologist who claims that evidence from the Human Genome Project proves we are all related to a single ancestral mother — the biblical Eve. No surprises there. If he were all there was to the evening's billing I would have stayed home. But along with the biologist the Creationist camp brought along the astronomer across from me. That is why I accepted the invitation to participate. I couldn't let the cosmological angle go by unchallenged. But rather than fight on the familiar turf of evidence for, or against, the Big Bang, I decided to use what I had been learning in writing this book and bring the debate to the other side of the field.

My new friend across the stage is finally done with his explanation. It is now my turn to rebut his points. I ask him what would happen if the Big Bang were disproved by new data. Would he give up his religion? After a long pause he says yes. I have a hard time with this. "You would give up being a Christian?" I ask disbelievingly. "Yes, I would," he replies again.

"You would give up on the grace of the Psalms? You would give up on the beauty of the Sermon on the Mount? You would give up on the vision of compassion and forgiveness that Jesus taught? You would give all that up because some spectroscopic data pushed a line on somebody's graph up rather than down?" There was another uncomfortable pause. "Yes," he says a bit sheepishly. "If the data showed the Big Bang was false I would consider my faith disproved." I didn't have the heart to push it any further, but I didn't believe him.

I still don't.

. . .

Throughout this book I have tried to argue that the old antagonisms between science and religion, between science and spiritual endeavor, have outworn their relevance. The scenes I described above typify that antagonism. On one side is the culture of science, which appears openly hostile to any discussion of how its domains touch those of human spiritual aspirations. Too often science appears relentless in its dismissal of spiritual life's relevance and its potential. By striking this cord we alienate too many people who are moved by the fruits of science's power and grand vision. On the other side are the religious literalists who insist that their particular vision of their particular scripture holds a monopoly on truth. The religious literalists' endless, losing battle with science appears buoyed only because they manage to win other battles in the arena of politics.

Somehow these poles seem to absorb the entirety of public discussion about science and religion. When I told a colleague I was writing a book on the subject he said, "Oh, so you will be writing a critique of

Intelligent Design." It is as if no one can even imagine there is anything else to talk about. Of course, there is the recent advent of Buddhist and Hindu perspectives in the debate. New Age wishful thinking has colored that well so effectively it is hard to separate the silliness from the serious discussion. In any case, those who argue for a science and Eastern religion connection share the same mistaken emphasis on results that haunts the Christian Fundamentalists. There has to be a better way.

In this book I have tried to sketch out a path to a different and more enlivened perspective on science and religion. I hope I have shown that another way to look at both domains exists and in that perspective one can see something rich and new and full of potential. I am hopeful that the connections I have explored here may prove useful in unshackling the debate about science and religion from the tired standard we have been chained to for so long. Changing the debate will free us to see these two great human efforts in the broadest possible terms and draw from both as we move forward with wisdom and balance. I believe that this change is not just possible but is, in fact, necessary if we are to navigate the challenges ahead.

Before we go any farther let me summarize the flow of the argument.

1. *Warfare is not the only way to tell the story of science and religion.* The history of science and religion is far more complicated and subtle than simply a narrative of deadly warfare between the two sides. The metaphor of war is a relatively recent creation. It was conditioned by historical events whose main players had quite specific reasons for telling the story in that way.

2. *The emphasis on results in science and religion is misguided and sterile.* The traditional public science/religion debate focuses on results. It is an endless comparison of what science says versus what some particular religion's doctrine holds. The emphasis on results misses a deeper and more fecund relationship between science and spiritual endeavor.

3. *Religious experience is more important than religious doctrine in thinking about connections with science.* The human phenomenon of religion begins with the very personal domain of religious experience. The emphasis on experience, rather than articles of faith or creed or dogma, provides a fundamentally different starting point for looking at what occurs at the root of spiritual life and its relation to science.

4. *Science, in its practice and its fruits, manifests hierophanies.* Religious experience is an encounter with the sacred character of being. The sacred is the opposite of the profane, "everyday," experience of life. Hierophanies can be identified as the location where the sacred erupts into our awareness, illuminating our experience of the world with a distinct quality of awe and reverence. Science provides us with hierophanies. It is a means to reveal the miracle that lies beneath every unconsidered moment. In this way science is a gateway to an experience of the sacred.

5. *Science functions as myth in providing hierophanies through sacred narratives of the cosmos and our place within it.* Our species' first attempt to make sense of the world was through myth. In myth's sacred narratives we draw closer to the world's unseen but deeply felt powers. Science now addresses these same issues through its own narratives and in doing so recalls and recovers myth's imperatives.

6. *Science's roots in myth reveal its living connections with spiritual endeavor.* The capacity for science to manifest hierophanies through its narratives has its roots in myth. Thus science is deeply rooted in the mythic tradition of human Being. Modern religious life can be followed back to the same root. Following this root draws science and spiritual endeavor into their proper, parallel, and active complementarity.

7. *Transcendent realities may or may not exist but are not necessary for science to be recognized as a means to apprehend the sacred.* Debates

about the nature of a Platonic realm in mathematics, or the existence of transhistorical archetypes in myth, will continue. The same is true of questions concerning the existence of the sacred as opposed to a sacred character of experience. Is there some eternal truth "out there" external to us, or is it all in our heads? These debates do not need to be resolved for us to begin developing a language that harmonizes science and spiritual endeavor. It is the open-ended quality of our lives, the fundamental mystery of our presence, that animates the effort in both domains.

8.   *The braiding of science and spiritual endeavor by means of their common roots in myth can support a global ethos for the application of science as we pass through the bottleneck of the next century.* The development of science has given humanity powers to alter its own habitat on a planetary scale. Thus, for the first time, we are forced to think of ourselves as a single species and act as such. Our abilities to marshal collective action will fail unless they are accompanied by narratives that provide meaning and illuminate sustaining values. These narratives can be found only by recognizing the link between science and our deepest sense of the sacred.

Each of the points above represents nothing more than the sketch of an idea. There is so much more work to be done. There is so much more that I would like to do in unpacking the tightly woven connections that emerge from these topics: the history of science and religion; the literature of religious studies scholars; the work of comparative mythologists; the vast domains of science itself in areas as broad as human origins and cosmology. There could be much more meat put on the skeleton provided by these chapters, and I hope to live long enough to carry forward some of that work. Still, the purpose of a skeleton is to provide an organism with structural stability. I am hopeful that the path outlined in this book contributes some resilient support for the ongoing efforts by many people to change the nature and temper of the dialogue on science and religion. And there are others. Writers such as Ursula Goodenough and Marcelo Gleiser, B. Alan Wallace, and Vic Mansfield have covered new ground thoughtfully and with great integrity. Even if I do not agree with all their claims, I see in their words a common desire to remain true to the practice of science while taking the promise of myth or spiritual endeavor seriously. There *is* something going on out there.

If nothing else, the path outlined in this book, the path leading from new historical research through ideas about religious experience and on to the braiding of science and mythology, shows us that other ways of talking about science and religion exist. Just traversing that path reveals something powerful. Issues in science and spiritual endeavor *can* be enthusiastically encountered without air being sucked from the room by tired carping over evolution versus Genesis.

SCIENCE AND RELIGION: THE MATRIX

> A good traveler has no fixed plans
> and is not intent upon arriving.
> A good artist lets his intuition
> lead him wherever it wants.
> A good scientist has freed himself of concepts
> and keeps his mind open to what is.
>
> Stephen Mitchell, *Tao Te Ching*

The problem with the science and religion debate is that we often don't know what we're talking about. Literally. Both science and religion have distinct and distinctly different faces as they relate to human life and culture. In the contentious science and religion debate it is often not clear which of these aspects are facing off against each other. Piet Hut, an astrophysicist at Princeton's Institute for Advanced Studies, has articulated three main facets that science and religion share. First there are the institutions of science and religion. These are the official organs of policy and standards: scientific journals, national academies, and Church hierarchies and their scriptural interpretations. Next are

Table 1. *Matrix of Interactions between Science and Religion*

Science

		Institutional	Cultural	Personal
Religion	Institutional	Religious Institutions' relationship to Scientific Institutions	Religious Institutions' relationship to Science in Culture	Religious Institutions' relationship to Personal Scientific Practice
	Cultural	Religion in Culture and its relationship to Scientific Institutions	Religion in Culture and its relationship to Science in Culture	Religion in Culture and its relationship to Personal Scientific Practice
	Personal	Personal Religious Experience and its relationship to Scientific Institutions	Personal Religious Experience and its relationship to Science in Culture	Personal Religious Experience and its relationship to Personal Scientific Practice

the cultural faces of science and religion. These are the places where each appears in the daily life of human culture. On the religion side there are, for example, services performed for deaths and weddings. On the science side there are classes in school, science museums, and popular magazines. Finally there is the personal level at which individuals encounter science or religion through some form of practice or interest. Hut defines a matrix of science and religion in terms of these different facets and the ways they relate to one another.

The bulk of debate in science and religion has focused squarely on institutions and their interactions: results published in scientific journals and the pronouncements of church or temple elders. In this book I have deliberately avoided those facets because they are the least relevant for the real issue at hand. When religion is not seen through its institutions it becomes spiritual endeavor. It becomes, in the best case, the heartfelt effort made in response to experiences of the world as sacred. When science is seen for more than its institutions the personal quality of a heartfelt response to the world can also be seen. Many scientists have their own experiences of the world's spiritual quality; often this is made manifest to them through their investigations. Thus what they explicitly discover and publish in sanctioned journals may be less important for the issues I am discussing than where their aspirations gain force and meaning.

One of the great difficulties of discussing science and religion is that this matrix of relationships is not even recognized. People think they are arguing over science and religion, but they are really debating about one small corner in a vast landscape of possibilities. I will engage in these institutional battles if needed. I will, for example, fight tooth and nail to keep religious literalists and their pointless antievolution bias out of school curricula. But it would be a tragic mistake to think that this battle defines the horizon of potential embraced by science and religion as a subject. In this book I have argued that the other squares in the matrix matter far more than the ones under the rubric "Institution." It is crucial to see that institutions, with their specific needs, exist at some distance from the place where the heat of the constant fire burns brightest. If we are willing to cross that distance we will eventually find ourselves returning home to find those most precious gifts that the universe, time, and evolution bestowed on us.

THE BURNING AT THE BEGINNING

> What a piece of work is a man! How noble in reason!
> how infinite in faculty! in form, in moving, how
> express and admirable! in action how like an angel!
> in apprehension how like a god! the beauty of the
> world! the paragon of animals!
>
> Shakespeare, *Hamlet*

It is all about aspiration. That, if anything, is the one point I hope will be taken from this book. So much of the public debate on science and religion focuses on big fat ideas: Faith versus Reason; Divinity versus Natural Law. Gasping for air in the heights of these institutionalized dichotomies, we miss the most beautiful, most hopeful part of both grand endeavors.

We are embodied social animals. We all find our deeply felt interior lives continually washing up on the shore of shared experience. The world we inhabit together is not composed of ideas but of vast blue skies harboring towering clouds and elaborate universes held by raindrops on summer leaves. All of it seems infused with a power that can be sensed but not grasped. All of it seems tender and poised to pass away. In those rare moments when we allow ourselves to notice, the world touches us with the utmost care and concern. From time to time its aching, excruciating beauty presses down on us and forces us to notice, forces us to see. Sometimes it is the sharp orange filigree in a stand of autumn trees that wakes us up. Sometimes it is the heavy silence of snow falling through the core of a street lamp's winter illumination. Sometimes it is stars, just stars, spilling across a cold night, spilling backward and forever as they cease to live on the surface of an upturned bowl and suddenly point exactly and explicitly to the infinite space below our feet. That remarkable world, that universe sensed beyond the senses, is the home of both Science and Religion. It is the place we all start and all return to.

It is the aspiration that matters. It is the aspiration, so keen and intense, that forms the most essential, enlivened tissue that connects science and religion. This yearning is born directly from our experience. It is born from our immersion in a world that is beautiful in its structure, nuanced in its forms, and elegant in its action. When we confront that world we can name it as nothing less than sacred. The aspiration fuels our efforts, quickens our resolve, and strengthens our will to know the source of the experiences more intimately. It may begin with a moment of reverie in a temple, church, or mountain glade. There, in a place found by accident on one's travels, a moment of supreme quiet appears and the self seems called to something beyond its petty concerns. Perhaps then it leads to study and prayer and contemplation. With this background it is called religious. The aspiration may just as easily begin in youth when a magnifying glass reveals the miracles resting on a leaf's surface. Then the sense of unseen wonders takes root and deepens with time. Perhaps it leads to years of work and a dedicated study of life and cells and genes. With this background it is called science. For the questions I have raised in this book the distinctions that follow these two examples are less important than the imperative that precedes them.

The aspiration to know the source of experience, the True and the Real, emanates from encounters with the world when it appears to us unmasked, unadorned, and unfiltered. That is when our encounters with the World, Universe, and Cosmos appear sacred. Regardless of what theorizing happens later, regardless of what individual doctrines evolve in attempts to describe what is True and what is Real, it is the original experience and the aspiration that grows from it that matter. The aspiration is original, and it is ancient. It is the fire, the constant fire, that defines our unique place in the hierarchies of life and being on this planet.

MYSTERIES, MEANING, AND SHARED DESTINY

> Here, where I am surrounded by an enormous landscape, which the winds move across as they come from the seas, here I feel that there is no one anywhere who can answer for you those questions and feelings which, in their depths, have a life of their own; for even the most articulate people are unable to help, since what words point to is so very delicate, is almost unsayable.
>   Live the questions now. Perhaps then, someday far in the future, you will gradually, without even noticing it, live your way into the answer.
>   Rainer Maria Rilke, *Letters to a Young Poet*

Throughout this book I have spoken of an essential mystery at the heart of human Being. I know there will be many scientists who will balk at this. It may seem to them to be essentially antirational or anti-intellectual. In a sense, I think such a criticism is not unfair. It points to the difference between what science does and what powers within us it draws from. I am passionately committed to science as a practice for seeing as deeply into the structure of the world as we can collectively reach. Still, some things just do not fall so easily into words and some questions have more meaning than their answers. Douglas Adams offers us a parable along these lines in the wonderful story of Deep Thought in *The Hitchhiker's Guide to the Galaxy*. Deep Thought is an uber-computer created by a hyper-intelligent race of beings to answer the ultimate question of Life, the Universe, and Everything. After millions of years of computation the answer turns out to be 42, an answer no one understands. In response an even bigger hyper-uber-computer and more millennia of waiting are required to figure out what the question means. This sums up the situation we face at the border between science and religion. The longing we feel for understanding, the aspiration driving science and spiritual endeavor, will never be satisfied by an equation written on a page or a proscription encoded in scripture. Our search for the truth, our call to the truth, is a living thing. We circle endlessly around it. Sometimes, perhaps, we gain a glimpse of its totality in the poetry of the Psalms or the Bhagavad Gita. Sometimes that glimpse comes through a derivation of Einstein's field equations. In all cases it is our lived experience that brings us to the boundaries of what is expressible and what cannot be fully framed in words or mathematics. How could it be otherwise? We are born and find ourselves in this incredible world of wonders. After some time, unknown to us beforehand, we pass, and then, who knows? It is strange and beautiful, delightfully crazy, and more than a little weird. Therein lies the mystery. Therein lies the opportunity.

Science and spiritual endeavor are both gateways. They are not the same. They are not equivalent. But together they arise from the same ancient location in our history and our being. Together they define what is best in us, and, if we can find the will and marshal the effort, they can guide our best efforts. Ultimately the fruit of science and spiritual endeavor is an understanding of human Being in its broadest context. It is a vision vast in its beauty, wide in its enduring mystery, and deep in its compassion. It is a vision that can only help, and help is exactly what we need as we face a common planetary destiny.

The stories that science tells elevate the individual into a collective with stewardship over the planet. A million years of evolution, with its gift of self-consciousness and the capacity to act as the eyes of the universe, place us in this strange position of cosmic teenagers. Where do we go from here? Do we make it to maturity, or do we destroy ourselves? The decision rests with our ability to hear the wisdom handed to us in our collective rites of passage — the wisdom embodied in our myths and now energized through myth's transference into science. If we act on the mysteries expressed in myth and the wonder revealed in science, then our horizons need not contract. The constant fire can burn for many millennia more as we grow wise enough to nourish its aspiration.

NOTES

PROLOGUE: HOUSE OF THE RISING SUN

1. For more on Newgrange, see Michael J. Kelly, *Newgrange: Archaeology, Art and Legend* (London: Thames & Hudson, 1995).

2. On this trip to Ireland I met a guy in a bar who told me that the Irish consider conversation a contact sport and that if you go to a party in Ireland you had better have a good story to tell even if it isn't true. In that spirit I take this opportunity to point out that this book is not a memoir but an argument for a new way of looking at science and spiritual endeavor. All of the stories I tell from my own experience are true, but I have cut and pasted and amplified where necessary to make my point.

3. This is clearly a male perspective. I am not sure if female scientists have a similar metaphor or perspective.

4. William James, *The Varieties of Religious Experience* (New York: Penguin Classics, 1982), 6.

5. Edward O. Wilson, *Consilience* (New York: Random House, 1998), 45.

6. There are serious and illuminating discussions in these arenas. The works of B. Alan Wallace and Piet Hut stand out as thoughtful encounters between science and contemplative practice.

7. Mircea Eliade, *Myths, Rites, Symbols: A Mircea Eliade Reader*, edited by Wendell C. Beane and William G. Doty (New York: Harper Colophon, 1959), 9.

8. Barbara Tuchman, *A Distant Mirror* (New York: Ballantine Books, 1987).

9. Ursula Goodenough, *The Sacred Depths of Nature* (New York: Oxford University Press, 1998).

1. THE ROOTS OF CONFLICT

1. Dorothea Singer, *Giordano Bruno: His Life and Thought* (New York: Greenwood Press, 1950), 180.

2. Antonio Calcagno, *Giordano Bruno and the Logic of Coincidence: Unity and Multiplicity in the Philosophical Thought of Giordano Bruno* (New York: Peter Lang, 1998), 25.

3. Robert Green Ingersoll, *The Works of Robert Ingersoll*, vol. 3 (New York: Ingersoll League, 1933).

4. Bruno Memorial Award, SETI League, www.setileague.org/awards/brunoquo.htm. Accessed June 22, 2007.

5. Calcagno, *Giordano Bruno and the Logic of Coincidence*, 19.

6. Calcagno, *Giordano Bruno and the Logic of Coincidence*, 9.

7. Richard Pogge, *The Folly of Giordano Bruno*, www.setileague.org/awards/brunoquo.htm. Accessed June 15, 2007.

8. Timothy Ferris, *Coming of Age in the Milky Way* (New York: Harper Collins, 2003), 65.

9. J. L. E. Dryer, *A History of Astronomy* (New York: Dover, 1963), 318.

10. John North, *The Norton History of Astronomy and Cosmology* (New York: Norton, 1995), 283.

11. Newberry Library, Glossary of Terms, www.newberry.org/k12maps/glossary/index.html. Accessed August 16, 2006.

12. Alister McGrath, *The Foundations of Dialogue in Science and Religion* (Oxford: Blackwell, 1998), 22.

13. Stout, quoted in MacGrath, *The Foundations of Dialogue*, 22.

14. McGrath, *The Foundations of Dialogue*, 23.

15. Percy Shelley, *The Complete Poetical Works of Shelley* (Boston: Houghton Mifflin, 1901).

16. S. F. Cannon, *Science in Culture: The Early Victorian Period* (New York: Science History Publications, 1978), 2.

17. McGrath, *The Foundations of Dialogue*, 21.

18. David C. Lindberg, "Exploring Science and Religion's Past," *Science and Theology News*, April 21, 2006. www.stnews.org/Books-2749.htm.

19. Lindberg, "Exploring Science and Religion's Past."

20. Bertrand Russell, *The History of Western Philosophy* (New York: Simon and Schuster, 1945), 528.

21. McGrath, *The Foundations of Dialogue*, 16.

22. McGrath, *The Foundations of Dialogue*, 16; E. Rosen, "Calvin's Attitudes towards Copernicus," *Journal of the History of Ideas* 21 (1960): 431.

23. Adam Shapiro points out that Draper and White were very selective in their reading of history. Each gave voice to groups that were concerned with specific religions. White's book was published at the height of "Nativism" and the immigration and education debates of his time. Draper explicitly targeted Catholicism. Adam Shapiro, interview by author, University of Chicago, March 13, 2007.

24. *Dictionary of Greek and Roman Biography and Mythology*, www.theoi.com/Titan/TitanPrometheus.html.

25. I have based my telling of the story on Padraic Colum, *The Golden Fleece and the Heroes Who Lived before Achillles* (New York: Simon and Schuster, 1949).

2. THE CONFLICT WE KNOW

1. Taken from Stephen Jay Gould's excellent essay "Nonoverlapping Magisteria," in his *Leonardo's Mountain of Clams and Diet of Worms* (New York: Three Rivers Press, 1998), 269.

2. The Fundamentals, *Welcome to the Fundamentals Homepage*, www.xmission.com/~fidelis/. Accessed March 20, 2007.

3. For the full text of the Butler Act and the law that repealed it, see *Tennessee Evolution Statutes*, www.law.umkc.edu/faculty/projects/ftrials/scopes/tennstat.htm. Accessed September 18, 2007.

4. This was due in part to the effect of the play, *Inherit the Wind* (1955), which was later made into a popular film (1960). See Edward J. Larson, *Summer for the Gods: The Scopes Trial and America's Continuing Debate over Science and Religion* (New York: Basic Books, 1997).

5. Barbara Forrest, *The Wedge at Work: Intelligent Design Creationism and Its Critics* (Cambridge, Mass.: MIT Press, 2001), 5.

6. Phillip E. Johnson, *Defeating Darwinism by Opening Minds* (Downers Grove, Ill.: InterVarsity Press, 1997), 92.

7. Quoted in Richard Dawkins, *The Blind Watchmaker* (New York: Norton, 1986), 4.

8. Dawkins, *The Blind Watchmaker*, 4.

9. Earth is now known to be approximately 5 billion years old.

10. National Academy of Sciences, *Science and Creationism: A View from the National Academy of Sciences* (Washington, D.C.: National Academy of Sciences Press, 1999), 10.

11. See, for example, John Rennie, "15 Answers to Creationist Nonsense," *Scientific American*, June 2002.

12. For an excellent overview of arguments for and against Intelligent Design, with sections by Behe and Wells, see "Intelligent Design?" *Natural History*, April 2002.

13. It has been argued that no scientific paper on Intelligent Design would be accepted for publication in an established peer-reviewed scientific journal because of the inherent bias of both editors and referees. I think there is some truth to this; many referees would be loath to open the floodgates. However, the history of science has repeatedly shown that if an idea can provide a better explanation of existing data and make predictions that can be tested quantitatively, then it will receive a hearing. Intelligent Design proponents have yet to clear the most basic hurdle of research. Their papers are rejected for the same reason that hundreds of other scientific papers are rejected: they do not meet the standards of good scholarship.

14. I do not count among the Silly the excellent works by B. Allen Wallace exploring physics and the role of the observer vis-à-vis contemplative practices. See, for example, B. Allen Wallace, *The Taboo of Subjectivity* (New York: Oxford University Press, 2000). For clearly reasoned and thoughtful arguments about modern physics and Christianity, see Ian G. Barbour, *Nature, Human Nature, and God* (Minneapolis: Ausburg Fortress, 2002).

15. Claims about Buddhist worldviews and physics must be distinguished from authentic research into the neuroscience and medical effects of meditative practices. While this research also has its critics it is at least producing results that can be judged on their own merits.

16. Quoted in Werner Heisenberg, *Physics and Beyond* (New York: Harper and Row, 1971), 206.

17. Jeffery Bub, *Interpreting the Quantum World* (Cambridge: Cambridge University Press, 1997).

18. Islam has played a role in the debate, but it was relatively minor over the past four hundred years. During the height of the Islamic Golden Age (ca. 750–1200), Islamic studies carried the development of science forward, but a resurgent Western Europe later reclaimed its preeminence in science. See

Taner Edis, "The World Designed by God," in *Science and Religion*, edited by Paul Kurtz (Amherst, N.Y.: Prometheus Books, 2003), 117.

3. SCIENCE AND THE SACRED

1. Larry Witham, *The Measure of God* (San Francisco: HarperCollins, 2005). I have drawn liberally from Witham's description of William James and his Gifford Lecture.

2. Witham, *The Measure of God*, 2.

3. Carol Zalesky, "William James The Varieties of Religious Experience," *First Things* 101 (March 2000): 60.

4. William James, *The Varieties of Religious Experience*, edited with an introduction by Martin E. Marty (New York: Penguin Classics, 1982), 31.

5. James, *Varieties*, xix. This quote is from the introduction to the Penguin edition by Martin E. Marty.

6. James, *Varieties*, 6.

7. James, *Varieties*, 30.

8. James, *Varieties*, 31.

9. The neurology of religious experience has become a domain of research, as discussed in chapter 2. See, for example, R. Joseph, ed., *NeuroTheology: Brain, Science, Spirituality, Religious Experience* (San Jose, Calif.: University Press, 2003).

10. James, *Varieties*, 14.

11. James, *Varieties*, 70.

12. Daniel Quinn, *Providence* (New York: Bantam, 1995), 63.

13. James, *Varieties*, 603.

14. James, *Varieties*, 58. Original emphasis.

15. James, *Varieties*, 610.

16. James, *Varieties*, 604.

17. Wayne Proudfoot, *Religious Experience* (Berkeley: University of California Press, 1985), 232.

18. Proudfoot, *Religious Experience*, xiv.

19. Proudfoot, *Religious Experience*, xv.

20. Rudolf Otto, *The Idea of the Holy* (Oxford: Oxford University Press, 1958), xxi.

21. Otto, *The Idea of the Holy*, 7.

22. Of course, the case can be made that "Whole Lotta Love" is a direct encounter with God.

23. Proudfoot, *Religious Experience*, 234.

24. Proudfoot, *Religious Experience*, 223.

25. Proudfoot, *Religious Experience*, 232.

26. Carsten Colpe, "The Sacred and the Profane," in *Encyclopedia of Religion*, edited by Mircea Eliade (New York: Macmillan, 1987), 12:511.

27. Colpe, "The Sacred and the Profane," 11.

28. Mircea Eliade, *The Sacred and the Profane: The Nature of Religion* (New York: Harcourt Brace, 1959), 8.

29. Eliade, *Sacred and Profane*, 10; my emphasis.

30. Eliade, *Sacred and Profane*, 10.

31. Eliade, *Sacred and Profane*, 11.

32. Eliade, *Sacred and Profane*, 11.

33. Andrew Wiget, "A Siberian Connection," www.nmsu.edu/~english/hc/hcsiberia.html; accessed September 19, 2007. For a description of Khanty practices, see Peter Jordan, *Material Culture and Sacred Landscape: The Anthropology of the Siberian Khanty* (New York: Rowman Altamira, 2003).

34. Eliade, *Sacred and Profane*, 14.

35. Eliade, *Sacred and Profane*, 13.

36. Ursula Goodenough, *The Sacred Depth of Nature* (New York: Oxford University Press, 1998), 12.

4. NOT THE GOD YOU PRAY TO

1. Jagdish Mehra and Helmut Rechenberg, *The Historical Development of Quantum Mechanics* (New York: Springer, 2001), xxxvi.

2. My description of Pauli's life comes from two sources: Charles P. Enz, *No Time to Be Brief* (New York: Oxford University Press, 2002); and the introduction to *Atom and Archetype: The Pauli/Jung Letters, 1932–1958*, edited by C. A. Mier (Princeton: Princeton University Press, 2001).

3. Enz, *No Time to Be Brief,* 211.

4. Carl Jung, *Psychology and Alchemy* (Princeton: Princeton University Press, 1980).

5. Ursula Goodenough, "Religious Naturalism Defined," http://religiousnaturalism.org/defs.html. Accessed September 10, 2006.

6. Elenor Bustin Mattes, *Myth for Moderns: Erwin Ramsdell Goodenough and Religious Studies in America, 1938–1955* (London: Scarecrow Press, 1997).

7. Ursula Goodenough, *The Sacred Depth of Nature* (New York: Oxford University Press, 1998), x.

8. Mattes, *Myth for Moderns*, xi; my emphasis.

9. Goodenough, *The Sacred Depth of Nature*, x.

10. This section was drawn from the work of Stephen Snobelen, professor of history and technology at University of Kings College, Halifax. Documents from the Newton Project, an international effort to research Newton's voluminous writings on theology, of which Snobelen is part, can be found at www .newtonproject.sussex.ac.uk. Accessed September 28, 2007.

11. The discussion on Maxwell and religion is taken from a number of sources, beginning with Ian Hutchinson, *James Clerk Maxwell and the Christian Proposition*, http://silas.psfc.mit.edu/Maxwell/maxwell.html; accessed September 23, 2007. See also Paul Theerman, "James Clerk Maxwell and Religion," *American Journal of Physics* 54, no. 312 (1986): 312–17.

12. Hutchinson, *James Clerk Maxwell and the Christian Proposition*.

13. Albert Einstein, *Ideas and Opinions, based on Mein Weltbild*, edited by Carl Seelig (New York: Bonanza Books, 1954), 8–11.

14. Alice Calaprice and Trevor Lipscombe, *Albert Einstein: A Biography* (New York: Greenwood Press, 2005), 5.

15. Max Jammer, *Einstein and Religion* (Princeton: Princeton University Press, 1999), 138.

16. Jammer, *Einstein and Religion*, 123.

17. Albert Einstein, *The World as I See It* (New York: Citadel Press, 2006), 27.

18. Jammer, *Einstein and Religion*, 97.

19. Einstein, *The World as I See It*, 5.

20. Einstein, *The World as I See It*, 28.

5. SCIENCE, MYTH, AND SACRED NARRATIVES

1. Edwin Krupp, *Beyond the Blue Horizon* (New York: HarperCollins, 1991), 17.

2. In my research on this subject one of the most well composed books on human origins that I found was Nicholas Wade, *Before the Dawn* (New York: Penguin Books, 2006). Wade, a science journalist for the *New York Times*, combines research on genetics, linguistics, and anthropology to paint a compelling story of our (continuing) evolution across the past 50,000 years. For a lovely graphic tour of human evolution, I recommend the Human Origins Program at the Smithsonian Institution, www.mnh.si.edu/anthro/humanorigins/aop/ aop_start.html.

3. *Becoming Human*, www.becominghuman.org/. Accessed September 20, 2006.

4. Robert J. Sternberg and James C. Kaufman, *The Evolution of Intelligence* (Philadelphia: Lawrence Erlbaum Associates, 2001), 71.

5. Russell Dale Guthrie, *The Nature of Neolithic Art* (Chicago: University of Chicago Press, 2005).

6. Much of this section comes from Karen Armstrong, *A Short History of Myth* (New York: Canongate, 2005). Note that Armstrong draws on Eliade quite heavily in her telling of the way myths changed during the course of the evolution of human culture.

7. Armstrong, *Short History of Myth*, 7.

8. Armstrong, *Short History of Myth*, 14.

9. Armstrong, *Short History of Myth*, 30.

10. Dennis D. Hughes, *Human Sacrifice in Ancient Greece* (Oxford: Routledge, 1991), 6.

11. Wade, *Before the Dawn*, 101.

12. Armstrong, *Short History of Myth*, 41.

13. Peter Ward and Donald Brownlee, *Rare Earth: Why Complex Life Is Uncommon in the Universe* (New York: Springer, 2003), 126.

14. Joseph Campbell, *Pathways to Bliss: Mythology and Personal Transformation* (New York: New World Library, 2004), 189.

15. Wendy Doniger, *A Very Strange Enchanted Boy, New York Times Book Review*, February 2, 1992, 7.

16. Joseph Campbell, *The Power of Myth* (New York: Doubleday, 1988).

17. Robert Alan Segal, *Theorizing about Myth* (Boston: University of Massachusetts Press, 1999), 19.

18. Krupp, *Beyond the Blue Horizon*, 17.

19. Robert Alan Segal, *A Very Short Introduction to Myth* (New York: Oxford University Press, 2004), 14.

20. Segal, *A Very Short Introduction to Myth*, 14.

21. Segal, *A Very Short Introduction to Myth*, 29.

22. Mircea Eliade, *Myth and Reality* (New York: Harper and Row, 1963), 5.

23. Mircea Eliade, *The Sacred and the Profane* (New York: Harcourt Brace, 1959), 205.

24. Eliade, *The Sacred and the Profane*, 205.

25. Segal, *Theorizing about Myth*, 19.

26. Segal, *Theorizing about Myth*, 30.

27. Leo Lionni, *Fredrick and His Friends* (New York: Random House, 1989).

28. Tiffany C., www.spaghettibookclub.org/review.php3?review_id = 301. Accessed September 30, 2007.

6. THE ORIGIN OF EVERYTHING

1. I have slightly modified this version of the story from a number of sources: Heinrich Robert Zimmer, *Myths and Symbols in Indian Art and Civilization* (Princeton: Princeton University Press, 1972), 3; Lone Jensen, www.vuu.org/sermons/ljo30223.htm, accessed September 10, 2007.

2. Dennis Danielson, ed., *The Book of the Cosmos* (Cambridge, Mass.: Helix Books, 2002), 265.

3. Bradley Carroll and Dale Ostlie, *Modern Astrophysics* (New York: Addison-Wesley, 1996), 93.

4. One of my favorite advanced textbooks on general relativity is Sean Carroll, *Spacetime and Geometry: An Introduction to General Relativity* (New York: Addison-Wesley, 2004).

5. A galaxy is a collection of 100 billion stars. All stars are part of galaxies, and galaxies are the basic "units" of structure in the universe.

6. A nice introduction to stories of scientific discoveries, including CMB, can be found in Franck Ashall, *Remarkable Discoveries!* (Cambridge: Cambridge University Press, 1994), 79.

7. Light can treated as both a particle, called a photon, and a traveling wave of electromagnetic energy. Microwaves are just another form of electromagnetic radiation. The electromagnetic spectrum includes everything from gamma rays to visual light to radio waves.

8. Helge S. Kragh, *Cosmology and Controversy* (Princeton: Princeton University Press, 1996), 219.

9. Actually they understood enough to push back to a few millionths of a second after the Big Bang, but I begin the story later. See Steven Weinberg, *The First Three Minutes: A Modern View of the Origin of the Universe* (New York: Basic Books, 1993).

10. The proton is the nucleus of the hydrogen atom. It does not count since thermonuclear fusion is not required to create a proton.

11. Andrew Liddle and David Lyth, *Cosmological Inflation and Large Scale Structure* (Cambridge: Cambridge University Press, 2000).

12. For more on dark matter and its cosmological implications, see Martin J. Rees, *Just Six Numbers: The Deep Forces That Shape the Universe* (New York: Basic Books, 2001).

13. It is not just the temperature but the way the temperature varies in CMB that was too uniform.

14. Liddle and Lyth, *Cosmological Inflation*, 4.

15. Alan H. Guth, *The Inflationary Universe: The Quest for a New Theory of Cosmic Origins* (New York: Basic Books, 1997).

16. For an engaging discussion of this debate, see Martin Rees, *Our Cosmic Habitat* (Princeton: Princeton University Press, 2003).

17. Marcelo Gleiser, *The Dancing Universe: From Creation Myths to the Big Bang* (New York: Penguin Group, 1998).

7. THE DELUGE THIS TIME

1. I adapted my telling of this myth from Stephanie Dalley, *Myths from Mesopotamia* (Oxford: Oxford University Press, 1998).

2. Roy Willis, *World Mythology* (New York: Owl Books, 1996), 63.

3. Gerrit Verschuur, *Impact: The Threat of Comets and Asteroids* (Oxford: Oxford University Press, 1996), 104.

4. My discussion of climate change science derives from a variety of sources; primary among them is Spencer R. Weart, *The Discovery of Global Warming* (Cambridge, Mass.: Harvard University Press, 2003). Weart is director of the Center for History of Physics of the American Institute of Physics, College Park, Maryland. His book is masterfully written and provides a broad discussion of the scientific issues as well as the historical pathways that led to their discovery. I draw liberally from the book and Weart's companion Web site and acknowledge my debt to his work. The companion Web site is Discovery of Climate Change, www.aip.org/history/exhibits/climate/.

5. Thule Air Force Base Home Page, *Camp Century*, www.thuleab.dk/index.php?option=com_content&task=view&id=27&Itemid=48.

6. John Cox, *Climate Crash* (Washington, D.C.: National Academies Press, 2005), 14.

7. Thomas Cronin, *Principles of Paleoclimatology* (New York: Columbia University Press, 1999), 414.

8. Roger Revelle, in *Scientific American* 247 (1982): 33–91.

9. For a technical discussion, see Satoh Masaki, *Atmospheric Circulation Dynamics and General Circulation Models* (New York: Springer, 2004).

10. Weart, *The Discovery of Global Warming*, 30.

11. The current state of the global model effort can be found in the latest executive summary of the IPCC preports or the detailed reviews created by the IPCC process. See Intergovernmental Panel on Climate Change, www .ipcc.ch/.ch/. Accessed June 19, 2007.

12. R. A. Warrick, E. M. Barrow, and T. M. Wigley, *Climate and Sea Level Change* (Cambridge: Cambridge University Press, 1993).

13. For an excellent popular account of the link between ice and climate, see Mark Bowen, *Thin Ice* (New York: Henry Holt, 2005).

14. Alan Dundes, *The Flood Myth* (Berkeley: University of California Press, 1988), 1.

15. Weart, *The Discovery of Global Warming*, 155.

16. Hans Kelsen, *The Principle of Retribution in Flood Myths*, edited by Alan Dundes (Berkeley: University of California Press, 1988), 125.

17. Dundes, *The Flood Myth*, 167.

18. Dundes, *The Flood Myth*, 167.

19. James Lovelock, *The Ages of Gaia* (New York: Norton, 1988).

8. MUSIC OF THE SPHERES

1. Walter Gratzer, *Eurekas and Euphorias: The Oxford Book of Scientific Anecdotes* (Oxford: Oxford University Press, 2004), 13; *The Houghton Mifflin Dictionary of Biography* (New York: Houghton Mifflin, 2004), 843.

2. W. H. Brock, *The Chemical Tree: A History of Chemistry* (New York: Norton, 2000), 264; Gerrylynn Roberts and Colin Russell, *Chemical History: Review of the Recent Literature* (London: Royal Society of Chemistry, 2005), 64.

3. We now know this conclusion is incorrect. Graeme Hunter, *Vital Forces: The Discovery of the Molecular Basis of Life* (San Diego: Elsevier, 2000).

4. A. E. Cavazos-Gaither, *Chemically Speaking: A Dictionary of Scientific Quotations* (Philadelphia: Institute of Physics Publishing, 2002), 158.

5. Eugene P. Wigner, "The Unreasonable Effectiveness of Mathematics in the Natural Sciences," in *Symmetries and Reflections* (Bloomington: Indiana University Press, 1967). Essay found at ew/cr.slu.edu/~srivasta/wigner.pdf. Subsequent quotations from Wigner are from this source.

6. Roger Penrose, *The Road to Reality* (New York: Knopf, 2005), 12.

7. George Lakoff and Rafael Nuñez, *Where Mathematics Comes From* (New York: Basic Books, 2001).

8. Lakoff and Nuñez, *Where Mathematics Comes From*, 366.

9. Carl Jung, *The Psychology of the Child Archetype;* quoted in Alan Dundes, *Sacred Narratives* (Berkeley: University of California Press, 1984), 245.

10. Dundes, *Sacred Narratives,* 245.

11. Jung, quoted in Dundes, *Sacred Narratives,* 245.

12. Joseph Campbell, *The Inner Reaches of Outer Space* (Novato, Calif.: New World Library, 2002), 69.

13. Robert Alan Segal, "The Romantic Appeal of Joseph Campbell," *Christian Century,* April 4, 1990, 332.

14. Segal, "The Romantic Appeal of Joseph Campbell," 332.

15. Werner, quoted in Wendy Doniger, *The Implied Spider* (New York: Columbia University Press, 1998), 138.

16. Doniger, *The Implied Spider,* 138.

17. Mircea Eliade, *Myths, Rites, Symbols: A Mircea Eliade Reader,* edited by Wendell C. Beane and William G. Doty (New York: Harper Colophon, 1976), 133.

18. Mircea Eliade, *Shamanism: Archaic Techniques of Ecstasy* (New York: Pantheon, 1964), xviii.

19. Eliade, *Myths, Rites, Symbols,* 138.

20. Doniger, *The Implied Spider,* 138.

21. Doniger, *The Implied Spider,* 145.

22. Lévi-Strauss, quoted in Doniger, *The Implied Spider,* 147.

23. Doniger, *The Implied Spider,* 147.

24. Doniger, *The Implied Spider,* 149.

25. For a scientific autobiography of Pauli, see Charles P. Enz, *No Time to Be Brief* (Oxford: Oxford University Press, 2002).

26. The correspondence between Pauli and Jung can be found in C. A. Meier, ed., *Atom and Archetype: The Pauli/Jung Letters, 1932–1958* (Princeton: Princeton University Press, 2001). A discussion of Pauli's ideas on science and archetypes can be found in Suzanne Giezer, *The Innermost Kernel* (New York: Springer, 2005).

27. Meier, *Atom and Archetype,* 179.

28. Meier, *Atom and Archetype,* 31.

29. Enz, *No Time to Be Brief,* 416.

30. Meier, *Atom and Archetype,* 203.

31. Meier, *Atom and Archetype,* 203.

32. Pauli, in Enz, *No Time to Be Brief,* 418.

33. Meier, *Atom and Archetype,* 37.

34. Ursula Goodenough, *The Sacred Depth of Nature* (New York: Oxford University Press, 1998), 224.

9. A NEED BORN OF FIRE

1. David A. Clary, *Rocket Man: Robert H. Goddard and the Birth of the Space Age* (New York: Hyperion, 2003).

2. Arthur C. Clarke, *The Promise of Space* (New York: Harper and Row, 1968), 15.

3. Clary, *Rocket Man*, 13.

4. Mike Gruntman, *Blazing the Trail: The Early History of Spacecraft and Rocketry* (Reston, Va.: American Institute of Aeronautics and Astronautics, 2004), 13.

5. Mircea Eliade, *The Myth of Eternal Return* (New York: Arkana, 1989).

6. James George Frazer, *The Illustrated Golden Bough* (London: George Rainbird, 1978).

7. James L. Christian, *Philosophy: An Introduction to the Art of Wondering* (Belmont, Calif.: Thompson/Wadsworth, 2005), 443.

8. David Scott and Alexei Lenov, *Two Sides of the Moon* (New York: St. Martin's Press, 2006), 81.

9. Robert Zubrin, "Getting Space Exploration Right," *The New Atlantis*, www.thenewatlantis.com/archive/8/zubrin.htm. Accessed June 18, 2007.

10. One of the best books on the road to the atomic bomb is Richard Rhodes, *The Making of the Atomic Bomb* (New York: Simon and Schuster, 1986).

11. Larry McCaffery, *Storming the Reality Studio: A Casebook of Cyberpunk and Postmodern Science* (Durham: Duke University Press, 1991).

12. Martin Rees, *Our Final Hour* (New York: Basic Books, 2003), 48.

13. George Musser, "The Climax of Humanity," *Scientific American* 243 (June 2005): 44.

14. Jared Diamond, *Collapse: How Societies Choose to Fail or Succeed* (New York: Penguin, 2006).

15. Edwin Schrodinger, *What Is Life* (Cambridge: Cambridge University Press, 1992), 117.

INDEX

Text:	Janson
Display:	Janson
Compositor:	BookMatters, Berkeley
Indexer:	Andrew Christenson
Printer and Binder:	Maple-Vail Book Manufacturing Group